Genetics and Genomics of the Brassicaceae

Genetics and Genomics of the Brassicaceae

Editors

Ryo Fujimoto
Yoshinobu Takada

MDPI • Basel • Beijing • Wuhan • Barcelona • Belgrade • Manchester • Tokyo • Cluj • Tianjin

Editors
Ryo Fujimoto
Graduate School of Agricultural Science
Kobe University
Kobe
Japan

Yoshinobu Takada
Graduate School of Life-Science
Tohoku University
Sendai
Japan

Editorial Office
MDPI
St. Alban-Anlage 66
4052 Basel, Switzerland

This is a reprint of articles from the Special Issue published online in the open access journal *Plants* (ISSN 2223-7747) (available at: www.mdpi.com/journal/plants/special_issues/genetics_genomics_brassicaceae).

For citation purposes, cite each article independently as indicated on the article page online and as indicated below:

LastName, A.A.; LastName, B.B.; LastName, C.C. Article Title. *Journal Name* **Year**, *Volume Number*, Page Range.

ISBN 978-3-0365-2657-7 (Hbk)
ISBN 978-3-0365-2656-0 (PDF)

© 2021 by the authors. Articles in this book are Open Access and distributed under the Creative Commons Attribution (CC BY) license, which allows users to download, copy and build upon published articles, as long as the author and publisher are properly credited, which ensures maximum dissemination and a wider impact of our publications.

The book as a whole is distributed by MDPI under the terms and conditions of the Creative Commons license CC BY-NC-ND.

Contents

Preface to "Genetics and Genomics of the Brassicaceae" vii

Hasan Mehraj, Ayasha Akter, Naomi Miyaji, Junji Miyazaki, Daniel J. Shea, Ryo Fujimoto and Md. Asad-ud Doullah
Genetics of Clubroot and Fusarium Wilt Disease Resistance in Brassica Vegetables: The Application of Marker Assisted Breeding for Disease Resistance
Reprinted from: *Plants* **2020**, *9*, 726, doi:10.3390/plants9060726 1

Naomi Miyaji, Mst Arjina Akter, Chizuko Suzukamo, Hasan Mehraj, Tomoe Shindo, Takeru Itabashi, Keiichi Okazaki, Motoki Shimizu, Makoto Kaji, Masahiko Katsumata, Elizabeth S. Dennis and Ryo Fujimoto
Development of a New DNA Marker for Fusarium Yellows Resistance in *Brassica rapa* Vegetables
Reprinted from: *Plants* **2021**, *10*, 1082, doi:10.3390/plants10061082 17

Rujia Zhang, Yiming Ren, Huiyuan Wu, Yu Yang, Mengguo Yuan, Haonan Liang and Changwei Zhang
Mapping of Genetic Locus for Leaf Trichome Formation in Chinese Cabbage Based on Bulked Segregant Analysis
Reprinted from: *Plants* **2021**, *10*, 771, doi:10.3390/plants10040771 29

Shengjuan Li, Charitha P. A. Jayasinghege, Jia Guo, Enhui Zhang, Xingli Wang and Zhongmin Xu
Comparative Transcriptomic Analysis of Gene Expression Inheritance Patterns Associated with Cabbage Head Heterosis
Reprinted from: *Plants* **2021**, *10*, 275, doi:10.3390/plants10020275 41

Yali Qiao, Xueqin Gao, Zeci Liu, Yue Wu, Linli Hu and Jihua Yu
Genome-Wide Identification and Analysis of *SRO* Gene Family in Chinese Cabbage (*Brassica rapa* L.)
Reprinted from: *Plants* **2020**, *9*, 1235, doi:10.3390/plants9091235 61

Haemyeong Jung, Seung Hee Jo, Won Yong Jung, Hyun Ji Park, Areum Lee, Jae Sun Moon, So Yoon Seong, Ju-Kon Kim, Youn-Sung Kim and Hye Sun Cho
Gibberellin Promotes Bolting and Flowering via the Floral Integrators *RsFT* and *RsSOC1-1* under Marginal Vernalization in Radish
Reprinted from: *Plants* **2020**, *9*, 594, doi:10.3390/plants9050594 75

Yoshinobu Takada, Atsuki Mihara, Yuhui He, Haolin Xie, Yusuke Ozaki, Hikari Nishida, Seongmin Hong, Yong-Pyo Lim, Seiji Takayama, Go Suzuki and Masao Watanabe
Genetic Diversity of Genes Controlling Unilateral Incompatibility in Japanese Cultivars of Chinese Cabbage
Reprinted from: *Plants* **2021**, *10*, 2467, doi:10.3390/plants10112467 97

Preface to "Genetics and Genomics of the Brassicaceae"

Brassicaceae is a diverse family of angiosperms containing 338 genera and 3709 species, including the model plant *Arabidopsis thaliana*. The genus *Brassica* includes many economically important crops providing nutrition as well as health-promoting substances. *Brassica rapa* L., including Chinese cabbage, pak choi, and turnip, and *Brassica oleracea* L., including cabbage, broccoli, cauliflower, and kohlrabi, show extreme morphological divergence due to selection by the plant breeders. Most cultivars of the *Brassica* vegetables are F_1 hybrids, and a breeding system was successfully established by effectively applying the phenomenon of heterosis/hybrid vigor, cytoplasmic male sterility, or self-incompatibility. *Brassica napus* comprises important oil seed crops, such as canola or rapeseed.

A famous diagram, Triangle of U, shows the genetic relationship between six species of the genus Brassica; three allotetraploid species, *Brassica juncea* L. (AABB), *Brassica napus* L. (AACC), and *Brassica carinata* L. (BBCC), were derived via hybridization between two diploid species, *B. rapa* (AA), *Brassica nigra* L. (BB), and *B. oleracea* (CC). Recently, whole-genome sequences have been determined in some species of Brassicaceae, and the detailed genetic relationships in allotetraploids featured in the U triangle have been revealed. In addition, resequencing in more than a hundred lines has shown genetic variation within a species. The basic information based on the reference genome sequence has greatly contributed to the advances in genetic and epigenetic analyses regarding various traits.

Ryo Fujimoto, Yoshinobu Takada
Editors

Review

Genetics of Clubroot and Fusarium Wilt Disease Resistance in Brassica Vegetables: The Application of Marker Assisted Breeding for Disease Resistance

Hasan Mehraj [1,*], Ayasha Akter [1,2], Naomi Miyaji [1], Junji Miyazaki [3], Daniel J. Shea [4,†], Ryo Fujimoto [1] and Md. Asad-ud Doullah [1,5,*]

1. Graduate School of Agricultural Science, Kobe University, Rokkodai, Nada-ku, Kobe 657-8501, Japan; 154a371a@stu.kobe-u.ac.jp or aakterhort@bau.edu.bd (A.A.); 162a318a@stu.kobe-u.ac.jp (N.M.); leo@people.kobe-u.ac.jp (R.F.)
2. Department of Horticulture, Bangladesh Agricultural University, Mymensing 2202, Bangladesh
3. Agriculture Victoria Research Division, Department of Jobs, Precincts and Regions, AgriBioscience, Bundoora, VIC 3083, Australia; junji.miyazaki@agriculture.vic.gov.au
4. Iwate Biotechnology Research Center, Narita, Kitakami, Iwate 024-0003, Japan; dshea30@bloomberg.net
5. Department of Plant Pathology and Seed Science, Faculty of Agriculture, Sylhet Agricultural University, Sylhet 3100, Bangladesh
* Correspondence: hmehraj34@stu.kobe-u.ac.jp (H.M.); asad.ppath@sau.ac.bd (M.A.-u.D.)
† Present Affiliation: Bloomberg LP, Chiyoda-ku, Tokyo 100-0005, Japan.

Received: 24 March 2020; Accepted: 27 May 2020; Published: 9 June 2020

Abstract: The genus Brassica contains important vegetable crops, which serve as a source of oil seed, condiments, and forages. However, their production is hampered by various diseases such as clubroot and Fusarium wilt, especially in Brassica vegetables. Soil-borne diseases are difficult to manage by traditional methods. Host resistance is an important tool for minimizing disease and many types of resistance (*R*) genes have been identified. More than 20 major clubroot (CR) disease-related loci have been identified in Brassica vegetables and several CR-resistant genes have been isolated by map-based cloning. Fusarium wilt resistant genes in Brassica vegetables have also been isolated. These isolated *R* genes encode the toll-interleukin-1 receptor/nucleotide-binding site/leucine-rice-repeat (TIR-NBS-LRR) protein. DNA markers that are linked with disease resistance allele have been successfully applied to improve disease resistance through marker-assisted selection (MAS). In this review, we focused on the recent status of identifying clubroot and Fusarium wilt *R* genes and the feasibility of using MAS for developing disease resistance cultivars in Brassica vegetables.

Keywords: clubroot; Fusarium wilt; *R* gene; quantitative trait locus; marker-assisted selection; Brassica

1. Introduction

The genus Brassica belongs to the family Brassicaceae (Cruciferae) containing 37 different species (http://www.theplantlist.org) and has great economic importance [1]. Three species, *Brassica rapa* L. (2n = 20, AA) and *Brassica oleracea* L. (2n = 18, CC) and its allotetraploid species, *Brassica napus* L. (2n = 38, AACC) are included in the genus Brassica and comprise commercially important vegetable and oilseed crops. *B. rapa* includes leafy vegetables such as Chinese cabbage (var. *pekinensis*), pak choi (var. *chinensis*), and komatsuna (var. *perviridis*), root vegetables such as turnip (var. *rapa*), and oilseed (var. *oleifera*). *B. oleracea* comprises commercially important vegetable crops with morphological variations such as cabbage (var. *capitata*), broccoli (var. *italica*), kale (var. *acephala*), kohlrabi (var. *gongylodes*), and cauliflower (var. *botrytis*). *B. napus* includes the oilseed crop, canola/rapeseed.

Various pathogens such as clubroot, Fusarium wilt, black rot, Sclerotinia stem rot, blackleg, white rust, downy mildew, white leaf spot, and turnip mosaic virus can infect Brassica crops [2,3]. Cultural, physical, biological, or chemical controls, or a combination of these controls, integrated pest management, are used for disease control. If plants have natural resistance against these pathogens, the dependence on these controls is reduced and is cost-effective. Thus, disease resistance is an important trait in plant breeding to prevent quality and yield losses.

The first tier of plant immunity is called pathogen-associated molecular pattern (PAMP)-triggered immunity (PTI) [4,5]. Plants recognize pathogens through the PAMPs by pattern recognition receptors (PRRs) [6] and this recognition leads to the activation of PTI. PTI induces the expression of defense genes such as the mitogen-associated protein kinase (MAPK) cascade or WRKY transcription factors [7,8]. In contrast, pathogens deliver virulence molecules called as effectors to suppress PTI [4]. The failure of PTI defense helps to activate an immune response called effector-triggered immunity (ETI), when plants recognize the effectors (Avr proteins) through disease resistance (R) proteins, an ETI is activated [5]. This recognition between R and Avr is termed 'gene-for-gene resistance' [9]. ETI is stronger against newly adapted pathogens in host plants than PTI [10]. R proteins contain nucleotide-binding (NB) and leucine rich repeat (LRR) domains, which are called NBS-LRR (nucleotide-binding site leucine-rich repeat) protein. NBS-LRR proteins are separated into two types by their N-terminus domain, either having a toll interleukin-1 receptor (TIR) domain (TIR-NBS-LRR protein) or coiled-coil (CC) domains (CC-NBS-LRR protein) [11–13]. In general, the LRR domain provides recognition specificity, the NB domain regulates activation, and the TIR domain regulates downstream signaling [5]. Besides this, some R genes also encode transmembrane receptor-like protein (RLPs), transmembrane receptor-like kinases (RLKs), cytoplasmic kinases (CKs), and proteins with atypical molecular motifs [4]. The constitution of R genes is different between monocotyledonous and dicotyledonous genomes. TIR-NBS-LRR genes are mostly absent in monocotyledons, while TIR-NBS-LRR genes are present in dicotyledons and usually more abundant than CC-NBS LRR genes [13]. The R genes have been comprehensively identified in several species of the genus Brassica [14–16].

In a practical sense, the successful deployment of a novel R gene into a crop depends on the identification of a positive phenotype, the identification of genetic markers for marker-assisted selection (MAS) breeding, and understanding of how the novel resistance will behave under different genetic backgrounds and pathogenic pressures in the field. Clubroot and Fusarium wilt are considered as devastating diseases, and they cause a significant yield loss of Brassica vegetables for many years over the world. Some clubroot-resistant lines are susceptible to the Fusarium wilt and vice versa. In this review, we focus on recent knowledge about R genes of clubroot and Fusarium wilt as several important R genes/quantitative trait loci (QTL) against these pathogens have been identified in Brassica vegetables. In addition, MAS has been used to improve the disease resistance, and several cultivars with higher resistance in Brassica vegetables have recently been developed. We will introduce recent information about R genes and the prospect of their possible utilization for Brassica breeding.

2. Infection Process of the Pathogens

2.1. Infection Process of Clubroot Pathogen P. brassicae

Clubroot is caused by the obligate parasite *Plasmodiophora brassicae* Woronin and is recognized as a major devastating disease in Brassicaceae that poses an emerging threat to Brassica crop production [17]. Clubroot disease was first reported in Russia in 1878 by Woronin and rapidly expanded to other countries like Europe, Brazil, South Africa, Australia, New Zealand, and China [17]. The infection of plants by *P. brassicae* is a two-phase process (Figure 1). The resting spores in soil germinate and the resultant zoospores then attack the plant's root hairs. The zoospores then grow into multi-nucleate plasmodia (primary plasmodia) within the root hairs. The plasmodia cleave the root tissues and form secondary zoospores. The secondary zoospores penetrate into the root cortical tissues in a process known as cortical infection [18,19]. This cortical infection induces abnormal growth by the development of secondary plasmodia inside the affected cell, and proliferation of the secondary plasmodia leads to the formation of distorted massive gall known as club [18–20]. During the development of the pathogen in the plant, these secondary zoospores are capable of infecting the same plant or adjacent plants, thus repeating the cycle. Secondary plasmodia develop into multinuclear plasmodia by a number of nuclear divisions, and further meiosis may appear before the formation of numerous resting spores within the diseased plant tissue [19,20]. Resting spores are released into soil by the decay of clubs and survive for many years in soil. The spores are spread field-to-field via drainage water and infected root debris. Clubroot inhibits nutrient and water transport, resulting in wilting and ultimately the death of the infected plant. It is difficult to control clubroot infection by any means except genetic resistance cultivars due to the longevity of the resting spores. Crop rotation by clubroot resistant cultivars can reduce 100% of the clubroot severity compared with the susceptible cultivars [21]. Practicing two or more years of crop rotation by clubroot resistant cultivars with clubroot host significantly reduces the resting spores in soil, which is near to complete eradication of clubroot [22]. The effective and sustainable clubroot management by clubroot resistant cultivars is now disclosed, and hints at the importance of resistant cultivars for clubroot management. On the other hand, the host-range of the pathogen is mostly restricted within Brassicaceae species [19,23,24].

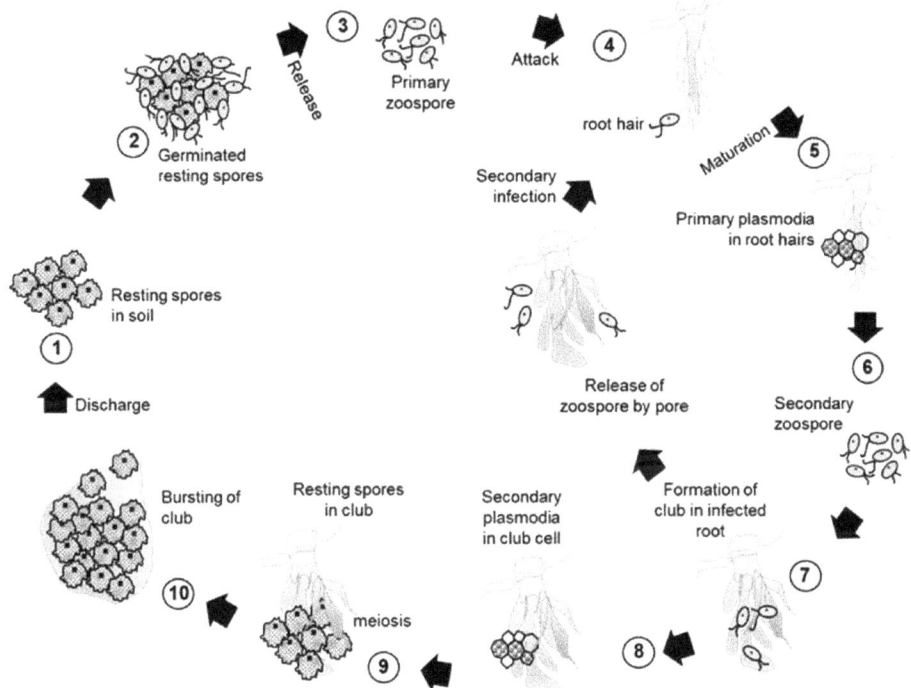

Figure 1. Infection process of clubroot disease caused by *Plasmodiophora brassicae*.

2.2. Infection Process of Fusarium Wilt Pathogen Foc/For

Yellowing or Fusarium wilt is caused by *Fusarium oxysporum f. sp. conglutinans/rapae* (Foc/For). Fusarium wilt disease was first reported in the USA, then in Japan and China, and has now been found almost all over the world [25,26]. The pathogen (Foc, *Fusarium oxysporum f. sp. conglutinans*)/For, *Fusarium oxysporum f. sp. rapae*) usually invades plants through their young roots, but can also invade through wounds in older roots [27,28]. This pathogen moves into and colonizes the xylem tissues, blocking vascular transport, leading to leaf yellowing, wilting, and defoliation, and in older plants, stunting and plant death [29,30]. The browning of vascular tissues can be observed in the stem and petiole of late-stage infected plants. It is a warm-weather disease and is active between 16 °C and 35 °C. The disease is more severe in warm conditions (above 24 °C) and not a problem in cool conditions [28,29]. The pathogen can survive in soil, seeds, and seedlings and can spread through water such as rain and flood [27,28] and remain for several years as resting spores in the soil. Two *forma specialis* (f. sp.) of *F. oxysporum* can cause disease in Brassicaceae. *Foc* causes disease in *B. oleracea* and *B. rapa* and *For* is specific to *B. rapa* [31]. Only two races in the *Foc*, race 1 and race 2, have been reported in the genus Brassica to date: race1 has been found worldwide and race 2 has only been found in USA and Russia [32].

3. Identification and Molecular Mechanism of Clubroot Resistant (CR) Genes

3.1. CR Loci in B. rapa

Clubroot disease resistance has been extensively studies in the genus Brassica. Several *CR* genes have been identified and mapped in *B. rapa*, *B. oleracea*, and other Brassica species [33]. In *B. rapa*, 18 major *CR* loci have been identified (Figure 2, Table 1); *Crr2* mapped on chromosome 1 [34], *CRc* and *CR* QTL, designated as *Rcr8*, on chromosome 2 [35,36], *Crr3*, *CRa*, *CRb*, *CRd*, *CRk*, *Rcr1*, *Rcr2*, and *Rcr4* on chromosome 3 [35–50], *CrrA5* on chromosome 5 [51], *Crr4* on chromosome 6 [52], *Crr1* (*Crr1a*, *Crr1b*), *Rcr9*, and *CRs* on chromosome 8 [36,44,53,54]. Most of the *CR* genes were identified through QTL mapping using a range of resistant sources based on molecular markers, genotyping-by-sequencing (GBS), or bulked segregant RNA sequencing (BSR-seq) strategies.

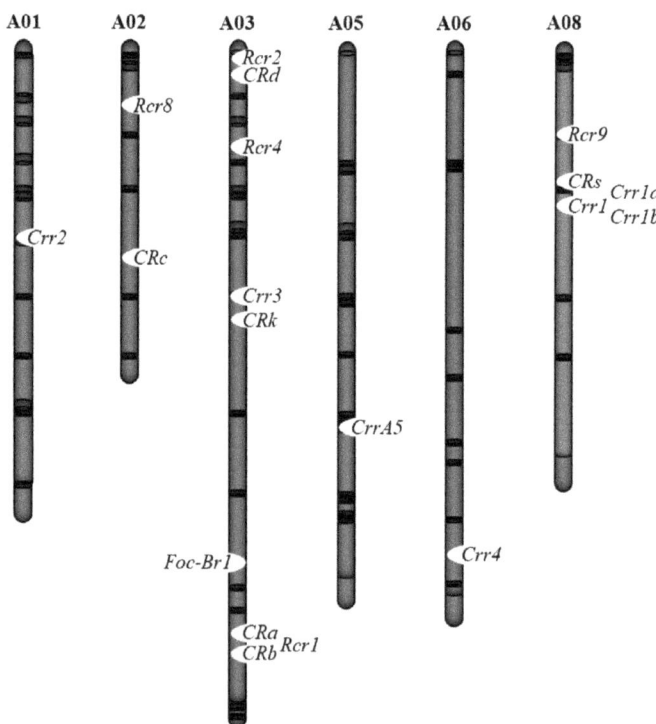

Figure 2. Chromosomal locations of clubroot resistant (CR) and Fusarium wilt resistant loci in *B. rapa*.

Table 1. CR loci reported on *B. rapa*.

QTLs	P/PR	Position	Linked Marker	Gene Source	References
Crr1	PR4	A08	BRMS-088	Turnip (G004-Siloga derived)	[34]
Crr1a	PR3,4	A08	BSA7	Turnip (G004-DH line)	[53]
Crr1b		A08	AT27		
Crr2	PR4	A01	BRMS-096	Turnip (G004-Siloga derived)	[34]
Crr3	PR3	A03	OPC11-2S	Turnip (Milan white)	[40,42]
Crr4	PR2,4	A06	WE24-1	Turnip (G004-Siloga derived)	[52]
CrrA5		A05	RAPD [1], SSR [2]	Chinese cabbage (Inbreed line 20-2ccl)	[51]
CRa	PR2	A03	HC352b-SCAR [3]	Chinese cabbage (DH line T136-8)	[39]
	PR2, P3			Chinese cabbage (CR Shinki)	[37,38]
	PR2,4,8		TCR09	Chinese cabbage (CR Shinki DH line, Akiriso)	[41]
CRb	P3	A03	KBrH059N21F	Chinese cabbage (CR Shinki)	[43]
	P3		B0902		[38,50]
	P4		KBrB085J21	Chinese cabbage (CR Shinki DH line)	[46]
CRc	PR2,4	A02	m6R	C9 (DH line of Debra)	[35]
CRd	PR4	A03		Chinese cabbage (Line 85–74)	[49]
CRk	PR2,4	A03	OPC11-2S	K 10 (DH line of CR Kanko)	[35]
CRs	P4	A08	SNP [4]	Chinese cabbage (cv. Akimeki)	[54]
Rcr1	P3	A03	SSR [2]	Flower Nabana (Pak choy)	[45]
	P2,5,6			Flower Nabana	[47]
Rcr2	P3	A03		Chinese cabbage (Jazz)	[48]
Rcr4	P2,3,5,6,8	A03			
Rcr8	P5X	A02	SNP [4]	Chinese cabbage (T19)	[36]
Rcr9	P5X	A08			

P, pathotypes; PR, physiological race of *P. brassicae*. [1] RAPD-Random Amplification of Polymorphic DNA, [2] SSR-Simple Sequence Repeat, [3] SCAR-Sequence Characterized Amplified Region, [4] SNP-Single Nucleotide Polymorphism.

The first *CR* gene was identified in the turnip cultivar Siloga using a doubled haploid (DH) population [55] and a dominant major gene *CRa* was mapped on chromosome 3. A candidate gene of *CRa* has been identified, and it encodes a TIR-NBS-LRR protein [37]. *Crr1a* and *CRb* genes have also been identified by map-based cloning [38,46,53]. *CRb* was isolated independently of *CRa*, but they were identical genes [37,38]. *Crr1a* encodes TIR-NBS-LRR [38].

Recently, proteomics in Chinese cabbage during response to *P. brassicae* infection identified differentially expressed proteins (DEPs) between the susceptible and resistant lines [56]. Gene ontology analysis using DEPs showed that the category of 'Glutathione transferase activity' was overrepresented, suggesting that glutathione transferase is responsible for protecting plants from disease [56].

3.2. CR Loci in B. oleracea

In contrast to *B. rapa*, no major *CR* genes or lines with strong resistance have been identified in *B. oleracea* [57]; only a few completely resistant accessions have been identified in *B. oleracea*. Genetic analysis of *CR* in *B. oleracea* was studied using diallel crossing methods or segregating populations. Only one major resistance gene, *Rcr7*, has been identified, and it might be located on chromosome 7 (LG 7) in cultivars, Tekila and Kilaherb of cabbage [57]. About fifty QTLs have been identified in the studies using different populations of *B. oleracea* (Table 2): three QTLs in broccoli [58], two in kale [59], two in cabbage [60], one in kale [61], three in kale [62], nine in kale [63], five in cabbage [64], three in cabbage using the GBS technique [65], and twenty-three QTLs in cabbage using single-nucleotide polymorphism (SNP) microarray technique [66]. The identification of several *CR* loci indicates that clubroot resistance in *B. oleracea* is controlled in a polygenic manner, confirming the complex genetic basis of the resistance, where a single resistance locus is not enough to confer sufficient resistance [67]. The comparison of these QTLs is currently impossible due to a lack of common molecular markers among different researchers and the use of different *CR* sources and races of pathogen [64].

Table 2. CR loci reported on *B. oleracea*.

QTLs	P/PR	Position	Linked Marker	Gene Source	References
Rcr7	P3,5X	LG7		Cabbage cv. Tekila and Kilaherb	[57]
3 QTLs	PR7	LG1	14a	Broccoli (CR-7)	[58]
		LG4	48		
		LG9	177b		
2 QTLs	ECD 16/31/31	-	OPL6-780, OPB11-740, OPA18-14900, OPA4-700, OPE20-1250, OPA1-1880, OPA16-510	Kale (C10)	[59]
Pb-3	ECD 16/3/30	LG3	4NE11a	Cabbage (Bindsachsener)	[60]
Pb-4		LG1	2NA8c		
1 QTL	PR2	LG3	WG6A1, WG1G5	Kale (K269)	[61]
QTL1		LG1	SCA02a2		
QTL3	PR2	LG3	SCB50b, SCB74c	Kale (K269)	[62]
QTL9		LG9	SOPT15a, SCA25		
Pb-Bo1		LG1	Ae05.8800, T2		
Pb-Bo2		LG2	PBB38a, r10.1200		
Pb-Bo3		LG3	Ae15.100, RGA8.450		
Pb-Bo4		LG4	ELI3.983, aa9.983		
Pb-Bo5a	P1,2,4,7	LG5	PBB7b, ae05.135	Kale (C10)	[63]
Pb-Bo5b		LG5	ELI3.115, a18.1400		
Pb-Bo8		LG8	C01.980, t16.500		
Pb-Bo9a		LG9	Aj16.570, W22B.400		
Pb-Bo9b		LG9	A04.1900, ae03.136		
Pb-Bo(Anju)1		LG2	KBrHo59L13		
Pb-Bo(Anju)2		LG2	CB10026	Cabbage (cv. Anju)	[64,67]
Pb-Bo(Anju)3	PR4	LG3	KBrB068C04		
Pb-Bo(Anju)4		LG7	KBrB089H07		
Pb-Bo(GC)1		LG5	CB10065		
2 QTLs	PR2	LG2		Cabbage (C1220)	[65]
1 QTL	PR9	LG3			
23 QTLs	PR4	-		Cabbage (GZ87)	[66]

P, pathotypes; PR, physiological race of *P. brassicae*; ECD, European Clubroot Differential set pathotype.

3.3. CR Loci other Brassica Species

In *B. napus*, the majority of CR identified genes are derived from *B. rapa* var. *rapifera* [57]. In *B. napus*, one dominant gene and more than 30 QTLs were identified (Table 3). Two QTLs, CR2a and CR2b, were identified using Rutabaga (cv. Wilhelmsburger) showing resistance to race 2 of *P. brassicae* [68]. A major gene, *Pb-Bn1*, mapped on chromosome A03 was reported first and two minor QTLs were mapped on linkage groups C02 and C09 [69]. Nineteen race-specific resistance QTLs were mapped on eight different chromosomes, including the A genome (A02, A03, A08, A09) and C genome (C03, C05, C06, C09) [70]. Besides this, five QTLs using a DH line of canola against pathotype 3 [71], and nine QTLs from different accession of oilseed rape were identified, seven of which were novel through integrative analysis [8]. They first applied genome-wide association study (GWAS) based on whole-genome SNP data to detect that nine QTLs and reported that these QTLs cover genes encoding TIR-NBS gene family [8]. Some resistance loci with one dominant and two recessive loci were identified [72], and one locus linked to *CRa* gene [73] and a genomic region on chromosome A08 carrying resistance to all five pathotypes, namely pathotypes 2, 3, 5, 6, and 8, were also identified [74]. This suggests that a single gene or a cluster of genes located in this genomic region is involved in the control of resistance to these pathotypes [74]. Recently, two major loci on chromosome A02 and A03 controlling resistance, and seven minor loci, were identified by a SNP association analysis [75].

A single dominant gene *Rcr6* was also identified on chromosome 3 of the B genome (B03) through BSR-Seq and further mapped with Kompetitive Allele Specific PCR (KASP) analysis in *Brassica nigra* lines PI 219,576 [33]. The authors declared that *Rcr6* was the first gene identified and mapped in the B genome of Brassica species. All of the *CR* genes found in the genus Brassica encode TIR-NBS-LRR proteins [57].

Table 3. CR loci reported on *B. napus* and *B. nigra*.

QTLs	PG/PR	Position	Process	Gene Source	References
B. napus					
CR2a	PR2	LG6	RFLP [2]	Rutabage (cv. Wilhelmsburger)	[68]
CR2b		LG1			
Pb-Bn1		A03			
1 QTL	P4,7	C02	RAPD [3]	Oilseed rape (cv. Darmor-bzh)	[69]
1 QTL		C09			
3 QTLs	SRSI	LG6 [1]	AFLP [4], SSR [5]	Canola (cv. Mendel)	[72]
19 QTLs	7 isolates with dissimilar P	A02, A03, A08, A09, C03, C05, C06, C09	AFLP [4], SSR [5]	Oilseed rape (cv. Boohmerwaldkohl and ECD04)	[70]
5 QTLs	P3,5,6,8	A03	SSR [5]/InDel [6]	Canola (cv. Mendel)	[71]
1 QTL	P3	A03	PCR-based marker	Canola (DH line 12-3, ECD04 derived)	[73]
1 QTL	P2,3,5,6,8	A08	SSR [5]	Rutabage (BF)	[74]
9 QTLs	P4	-		Oilseed rape (different accession)	[8]
2 QTLs	ECD 17/31/31	A02, A03	SNP [7]	Oilseed rape	[75]
B. nigra					
Rcr6	P3	B03		Accession PI 219,576 (parental line)	[33]

P, pathotypes; PR, physiological race of *P. brassicae*; SRSI, Single Resting Spore Isolate of *P. brassicae*; ECD, European Clubroot Differential set pathotypes. [1] Dominant locus (with two recessive loci), [2] Restriction Fragment Length Polymorphism, [3] Random Amplification of Polymorphic DNA, [4] Amplified Fragment Length Polymorphism, [5] Simple Sequence Repeat, [6] Insertion-Deletion, [7] Single Nucleotide Polymorphism.

4. Identification and Molecular Mechanism of Fusarium Wilt Resistance Genes

Two types of resistance (Type A and Type B) in *B. oleracea* have been reported against Fusarium wilt [76]. Type A resistance is controlled by a single dominant gene and is stable at temperatures higher than 24 °C where Type B is polygenic and becomes unstable at temperatures above 24 °C [27,76,77]. Type A resistance is controlled by a single dominant gene against race 1 in *B. rapa* and *B. oleracea* and has been studied extensively in recent years (Figure 3, Table 4) [29,30,32,78–80].

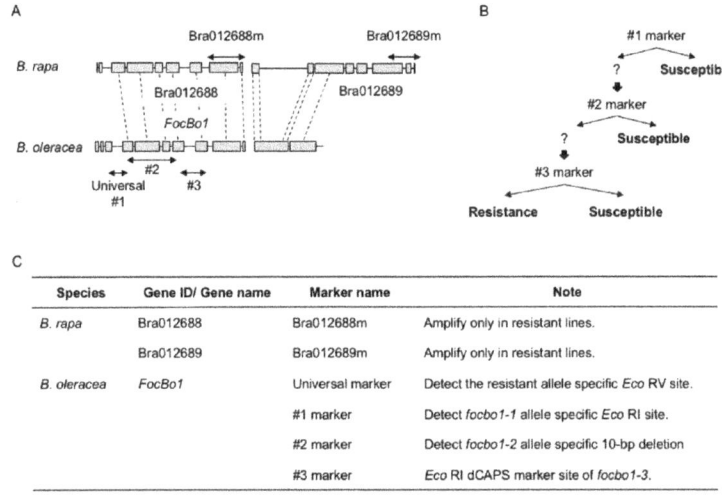

Figure 3. Schematic view of the alignment of resistance genes of Fusarium wilt disease. (**A**). DNA marker positions of resistance genes in *B. rapa* and *B. oleracea*. Arrows indicate marker positions. (**B**). Scheme of marker assisted selection in *B. oleracea*. (**C**). DNA marker list for marker assisted selection in *B. rapa* and *B. oleracea*.

Table 4. Loci of resistance gene to *Foc* reported in Brassica species.

QTLs	Position	Linked Marker/Process	Gene Source	References
B. rapa				
Foc-Br1a	A03	Bra012688m	Chinese cabbage (F_2 population)	[81]
Foc-Br1b		Bra012689m		
B. oleracea				
FOC	C06	InDel marker: M10 and A1	Cabbage (DH lines)	[78]
FOC1[1]	C06	InDel marker: Bol037156 and Bol037158	Cabbage (DH line and F_2 population)	[32]
QTL1	C04	SSR marker: KBrS003O1N10	Cabbage (AnjuP01): F_2 population	[29]
QTL2 (Foc-Bo1)[2]	C07			
Foc-Bo1[1] (SDG)	C07	InDel marker: BoInd 2 and BoInd 11	Cabbage (AnjuP01): Recombinant F_2 population	[30]

[1] Single dominant gene, [2] major QTL.

In *B. rapa*, transcriptome analysis was performed using resistant and susceptible lines. The differentially expressed *R* genes were identified and seven dominant DNA markers at *R* genes were developed. Two dominant DNA markers on Bra012688 and Bra012689 were completely linked to the resistance phenotype by an inoculation test, indicating that these two genes are candidates for Fusarium wilt resistance genes in *B. rapa* (Figure 3, Table 4). These two genes encode TIR-NBS-LRR proteins [81]. Dominant DNA markers, Bra012688m and Bra012689m, were applied to Chinese cabbage inbred lines and confirmed close linkage to the Fusarium wilt resistant phenotype [82]. Furthermore, the transcriptome profiles following *Foc* inoculation between Fusarium wilt-resistant and -susceptible lines in *B. rapa* were compared and differentially expressed genes were identified [79]. These genes may be responsible for the resistance mechanism to *Foc* [79]. Differentially expressed genes between *B. rapa* and *Arabidopsis thaliana* after *Foc* inoculation at the same time point were compared and up-regulated genes related to defense response were identified [79], that may be candidates for conferring resistance against *Foc*.

Recently, Type A resistance has been mapped and molecular markers have been developed in *B. oleracea* [29,30,32]. The Fusarium wilt resistance gene, *FocBo1*, was mapped on chromosome 7 by both segregation testing and QTL analysis, and the closest simple sequence repeat (SSR) marker KBrS003O1N10 was developed [29]. One minor QTL was also detected on chromosome 4. In a previous study, the resistance gene on chromosome 6 of cabbage was linked to two insertion/deletion (InDel) markers: M10 and A1 [78]. Later, it was shown that the resistance of Fusarium wilt was controlled by a single dominant gene based on the segregation ratio of two populations (resistant inbred line, 99–77 and highly susceptible line, 99–91). Two *R* genes in the target region, re-Bol037156 and re-Bol0371578, were predicted as resistance genes, and re-Bol037156 gene, which encodes a putative TIR-NBS-LRR type R protein, has highly similar sequences among the resistant lines [31]. *FocBo1* locus was identified on chromosome 7 and this locus was fine-mapped by using 139 recombinant F_2 plants derived from resistant cabbage (AnjuP01) and susceptible broccoli (GCP04) DH lines [30]. The *FocBo1* gene was shown by fine mapping to be an orthologous gene of Bra012688 in Chinese cabbage [30].

The proteome of xylem sap of the non-infected and *Foc* infected plants in both resistant and susceptible cabbage cultivars was also investigated using liquid chromatography-tandem mass spectrometry (LC-MS/MS) after the in-solution digestion of xylem sap proteins [83]. Twenty-five proteins in the infected xylem sap were found and ten of them were cysteine-containing secreted small proteins, suggesting that they are candidates for virulence and/or avirulence effectors. The transcriptome profiling of resistance to *Foc* in cabbage roots were also analyzed [26], where 885 differentially expressed genes (DEGs) were identified between infected and control samples at 4, 12, 24, and 48 h after inoculation. Some genes involved in Salicylic acid (SA)-dependent systemic acquired resistance (SAR), ethylene (ET)-, jasmonic acid (JA)-mediated, and the lignin biosynthesis pathways showed differential expression; the authors discussed the possibility that DEGs involved in these pathways may play important roles in resistance against *Foc* inoculation [26].

5. Resistant Breeding, Gene Accumulation, and MAS

MAS is an indirect selection process where a trait of interest is selected based on a marker (morphological, biochemical, or DNA/RNA variation) linked to that trait. Selecting individuals with disease resistance using MAS involves identifying a marker allele that is linked to disease resistance rather than to the level of disease resistance. There are several types of DNA markers that have been used to identify disease resistance genes [32,41,43,46,59,78,82,84].

The complexity of plant–pathogen interaction is a problematic in the case of CR breeding due to the appearance of multiple races of the pathogen [85]. Combinations of different CR genes exhibit higher resistance to the disease [62,67,86]. Though CR cultivars have been used widely for major production areas, field isolates of *P. brassicae* show variation, and different resistance sources from either *B. rapa* or *B. oleracea* vegetables were attained by *P. brassicae*. This suggests a serious risk that a resistance gene can be overcome by pathogen variants [3]. For example, seven CR canola cultivars were characterized for virulence in 106 *P. brassicae* population, and 61 of 106 *P. brassicae* population overcame the resistance in at least one of the seven CR cultivars [87]. There are many reports that CR genes show different reactions against the variable virulence of *P. brassicae* [34,36,44,48,61–63,66,74], but heterozygous CR loci are less resistant than the homozygous state [18]. *B. rapa* possesses several major CR loci (Table 1), which may confer differential (pathotype-specific) resistance to particular isolates of *P. brassicae*, and sometimes this may have a large effect on resistance [34,52,85,88]. The NARO Institute of Vegetable and Tea Science (NIVTS) has developed a high CR Chinese cabbage cultivar, 'Akimeki', by the accumulation of *Crr1*, *Crr2*, and *CRb* genes. It was proven that the accumulation of CR genes through MAS strengthened resistance and, consequently, it can be resistant to the multiple races of *P. brassicae* in *B. rapa*. Three CR genes, *CRa*, *CRk*, and *CRc*, were accumulated in Chinese cabbage through MAS [85] and the homozygous lines for the CR genes exhibited exceedingly high resistance against all six field isolates of *P. brassicae*. The effect of accumulation of different CR genes could be controlled by the dose-dependent accumulation of CR proteins [53,89]. In *B. oleracea*, resistance in genotypes has generally been identified less frequently than in the genotypes of *B. rapa* and the level of resistance is low [90]. This might be due to the polygenic nature of resistance in *B. oleracea* [67]. *B. oleracea* progeny were developed by accumulating major and minor QTLs to evaluate its effectiveness to the clubroot disease [64]. Three QTLs in the F_2/F_3 population from the cross between cabbage and kale line K269 were identified [62]. The accumulation of those three CR genes showed broad resistance to three isolates. It was observed that only one major QTL PbBo(Anju)1 showed moderate resistance, whereas three minor QTLs without the major one showed distinct susceptibility [64]. Later, it was proven that PbBo(Anju)1 and three minor QTLs PbBo(Anju)2, PbBo(Anju)4, and PbBo(GC)1 play a critical role in the acquisition of resistance to clubroot disease [67,86]. Here, *PbBo(Anju)1* plays a crucial role in the expression of clubroot resistance, and pyramiding minor CR genes are also essential for achieving higher resistance [67,86]. Their effectiveness was verified for controlling disease involving various isolates of *P. brassicae* [67]. Recently, two CR genes, *CRb* and *PbBa8.1*, were combined through MAS and CR homozygous lines in developed *B. napus*. The homozygous lines demonstrated a higher resistance than the heterozygous lines [91].

The Type A resistance to Fusarium wilt disease controlled by a single dominant gene has been successfully mapped and molecular markers have been developed: SSR marker KBrS003O1N10 [29], InDel markers M10 and A1 [78], Indel markers Bra012688m and Bra012689m [81,82], and DNA marker sets [30,80,84], which are used to generate a series of resistance cultivars (Figure 3).

Breeding cultivars that have resistance to both clubroot and Fusarium wilt is desired. However, inoculation tests against multiple pathogens or multiple races are difficult to perform on the same individual plant. Thus, DNA marker-based selection is useful for the identification of plants that have one Fusarium wilt resistance gene and multiple clubroot resistance genes. Furthermore, it is necessary to confirm whether these resistance genes are linked. In *B. rapa*, a Fusarium wilt resistance gene is located on chromosome 3, and *CRa/CRb*, *Crr3*, and *CRk* are located near this Fusarium wilt resistance gene. The *CRa/CRb* gene is the closest, being approximately 2 Mb in physical distance to the Fusarium wilt resistance gene (Figure 2). Since recombination between these two genes can occur [82], it is possible to inherit both resistance genes. In *B. oleracea*, a Fusarium wilt resistance gene is located on chromosome 7, and there is a minor QTL for clubroot resistance, PbBo(Anju)4, nearby this Fusarium wilt resistance gene. However, these loci are not completely linked to each other [81,83,84]. Therefore, it is possible to have both resistance genes. In *B. napus*, the association between susceptibility to Fusarium wilt and clubroot resistance against pathotype 3 was found, and these two resistance genes are located about 10 cM apart [92]. However, recombination between these two genes has been reported [92], suggesting that it is possible to inherit both resistance genes and identify them by DNA marker-based selection.

From the results from various researchers, it has been demonstrated that the DNA markers developed can select for the genes that are required for the acquisition of resistance, and these markers could be a powerful tool for resistance breeding in Brassica species. The novel breeding method developed can reinforce resistance by pyramiding *R* genes through MAS. For the genetic accumulation of *R* genes corresponding to wide pathogenicity, MAS is indispensable because it allows a precise identification of how many *R* genes are involved in a cultivar, and can monitor the accumulation of *R* genes in the progeny in the breeding program. To increase the durability of resistant cultivars to a broader spectrum of pathogen races, the combination of different *R* genes into a single line will be indispensable.

6. Conclusions

Brassica production is hampered by various diseases, especially clubroot and Fusarium wilt. Many types of *R* genes/QTLs have been identified in Brassica against the diseases and are being used for the improvement of resistance in cultivars. In case of clubroot disease, a total of 18 major *CR* loci have been identified in *B. rapa*, whereas only one major *CR* locus (*Rcr7*) and about 50 QTLs were detected in *B. oleracea*. Moreover, one locus (*Pb-Bn1*) on the A genome with more than 30 QTLs in *B. napus* and one locus (*Rcr6*) on the B genome in *B. nigra* were also identified. Several types of DNA markers that are linked with disease resistance allele have been developed, and they have been used for MAS. However, when there are several pathotypes, it is necessary to match effective *R* genes with a specific pathotypes and develop the DNA markers. The accumulation of *CR* genes corresponding to a wide pathogenicity will be important for breeding resistant cultivars.

A single type A dominant locus (*Foc-1*) was identified in *B. rapa* and *B. oleracea*, and several DNA markers have been developed.

R genes found from both diseases mostly encode a putative TIR-NBS-LRR. Understanding how plants cope with exposure to multiple pathogens such as *P. brassicae* and *Foc* will be important in breeding cultivars with multiple disease resistance.

Author Contributions: Conceptualization, H.M., R.F., and M.A.-u.D.; writing—original draft preparation, H.M., A.A., N.M., J.M., D.J.S., R.F., and M.A.-u.D.; supervision, wrote the manuscript. R.F.; funding acquisition, R.F. and M.A.u.D. All authors have read and agreed to the published version of the manuscript.

Funding: This work was funded by grants from Project of the NARO Bio-oriented Technology Research Advancement Institution (Research program on development of innovation technology) and International Research Fellow of JSPS (Invitation Fellowships for Research in Japan (Long-term)).

Acknowledgments: We thank Kenji Osabe and Elizabeth S. Dennis for their helpful comments and manuscript editing.

Conflicts of Interest: The authors declare no conflict of interest.

References

1. Dixon, G.R. Vegetable Brassicas and related crucifers. In *Crop Production Science in Horticulture Series*; Atherton, J., Ed.; CABI: Wallingford, UK, 2007; Volume 14, p. 327.
2. Lv, H.; Fang, Z.; Yang, L.; Zhang, Y.; Wang, Y. An update on the arsenal: Mining resistance genes for disease management of *Brassica* crops in the genomic era. *Hortic. Res.* **2020**, *7*, 34. [CrossRef] [PubMed]

3. Neik, T.X.; Barbetti, M.J.; Batley, J. Current status and challenges in identifying disease resistance genes in *Brassica napus*. *Front. Plant Sci.* **2017**, *8*, 1788. [CrossRef] [PubMed]
4. Jones, J.D.; Dangl, J.L. The plant immune system. *Nature* **2006**, *444*, 323–329. [CrossRef] [PubMed]
5. Dodds, P.N.; Rathjen, J.P. Plant immunity: Towards an integrated view of plant-pathogen interactions. *Nat. Rev. Genet.* **2010**, *11*, 539–548. [CrossRef]
6. Boller, T.; Felix, G. A renaissance of elicitors: Perception of microbe-associated molecular patterns and danger signals by pattern-recognition receptors. *Annu. Rev. Plant Biol.* **2009**, *60*, 379–406. [CrossRef]
7. Bigeard, J.; Colcombet, J.; Hirt, H. Signaling mechanism in pattern-triggered immunity (PTI). *Mol. Plant* **2015**, *8*, 521–539. [CrossRef]
8. Li, L.; Luo, Y.; Chen, B.; Xu, K.; Zhang, F.; Li, H.; Huang, Q.; Xiao, X.; Zhang, T.; Hu, J.; et al. A genome-wide association study reveals new loci for resistance to clubroot disease in *Brassica napus*. *Front. Plant Sci.* **2016**, *7*, 1483. [CrossRef]
9. Dangl, J.L.; Jones, J.D. Plant pathogens and integrated defence responses to infection. *Nature* **2001**, *411*, 826–833. [CrossRef]
10. Noman, A.; Aqeel, M.; Lou, Y. PRRs and NB-LRRs: From signal perception to activation of plant innate immunity. *Int. J. Mol. Sci.* **2019**, *20*, 1882. [CrossRef]
11. Meyers, B.C.; Kozik, A.; Griego, A.; Kuang, H.; Michelmore, R.W. Genome-wide analysis of NBS-LRR-encoding genes in Arabidopsis. *Plant Cell* **2003**, *15*, 809–834. [CrossRef]
12. Joshi, R.K.; Nayak, S. Functional characterization and signal transduction ability of nucleotide-binding site-leucine-rich repeat resistance genes in plant. *Genet. Mol. Res.* **2011**, *10*, 2637–2652. [CrossRef]
13. Marone, D.; Russo, M.A.; Laido, G.; De Leonardis, A.M.; Mastrangelo, A.M. Plant nucleotide binding site-leucine-rich repeat (NBS-LRR) gene: Active guardians in host defense responses. *Int. J. Mol. Sci.* **2013**, *14*, 7302–7326. [CrossRef]
14. Yu, J.; Tehrim, S.; Zhang, F.; Tong, C.; Huang, J.; Cheng, X.; Dong, C.; Zhou, Y.; Qin, R.; Hua, W.; et al. Genome-wide comparative analysis of NBS-encoding genes between *Brassica* species and *Arabidopsis thaliana*. *BMC Genom.* **2014**, *15*, 3. [CrossRef]
15. Bayer, P.E.; Golicz, A.A.; Tirnaz, S.; Chan, C.K.; Edwards, D.; Batley, J. Variation in abundance of predicted resistance genes in the *Brassica oleracea* pangenome. *Plant Biotechnol. J.* **2019**, *17*, 789–800. [CrossRef]
16. Dolatabadian, A.; Bayer, P.E.; Tirnaz, S.; Hurgobin, B.; Edwards, D.; Batley, J. Characterization of disease resistance genes in the *Brassica napus* pangenome reveals significant structural variation. *Plant Biotechnol. J.* **2020**, *18*, 969–982. [CrossRef]
17. Perez-Lopez, E.; Waldner, M.; Hossain, M.; Kusalik, A.J.; Wei, Y.; Bonham-Smith, P.C.; Todd, C.D. Identification of *Plasmodiophora brassicae* effectors—A challenging goal. *Virulence* **2018**, *9*, 1344–1353. [CrossRef]
18. Hirai, M. Genetic analysis of clubroot resistance in *Brassica* crops. *Breed. Sci.* **2006**, *56*, 223–229. [CrossRef]
19. Kageyama, K.; Asano, T. Life cycle of *Plasmodiophora brassicae*. *J. Plant Growth Regul.* **2009**, *28*, 203–211. [CrossRef]
20. Schwelm, A.; Fogelqvist, J.; Knaust, A.; Jülke, S.; Lilja, T.; Bonilla-Rosso, G.; Karlsson, M.; Shevchenko, A.; Dhandapani, V.; Choi, S.R.; et al. The *Plasmodiophora brassicae* genome reveals insights in its life cycle and ancestry of chitin synthases. *Sci. Rep.* **2015**, *5*, 11153. [CrossRef]
21. Hwang, S.F.; Ahmed, H.U.; Zhou, Q.; Fu, H.; Turnbull, G.D.; Fredua-Agyeman, R.; Strelkov, S.E.; Gossen, B.D.; Peng, G. Influence of resistant cultivars and crop intervals on clubroot of canola. *Can. J. Plant Sci.* **2019**, *99*, 862–872. [CrossRef]
22. Ernst, T.W.; Kher, S.; Stanton, D.; Rennie, D.C.; Hwang, S.F.; Strelkov, S.E. *Plasmodiophora brassicae* resting spore dynamics in clubroot resistant canola (*Brassica napus*) cropping systems. *Plant Pathol.* **2019**, *68*, 399–408. [CrossRef]
23. Rolfe, S.A.; Strelkov, S.E.; Links, M.G.; Clarke, W.E.; Robinson, S.J.; Djavaheri, M.; Malinowski, R.; Haddadi, P.; Kagale, S.; Parkin, I.A.P.; et al. The compact genome of the plant pathogen *Plasmodiophora brassicae* is adapted to intracellular interactions with host *Brassica* spp. *BMC Genom.* **2016**, *17*, 272. [CrossRef] [PubMed]
24. Wagner, G.; Laperche, A.; Lariagon, C.; Marnet, N.; Renault, D.; Guitton, Y.; Bouchereau, A.; Delourme, R.; Manzanares-Dauleux, M.J.; Gravot, A. Quantitative resistance to clubroot deconvoluted into QTL-specific metabolic modules. *J. Exp. Bot.* **2019**, *70*, 5375–5390. [CrossRef] [PubMed]
25. Smith, E.F. The fungus infection of agricultural soils in the United States. *Sci. Am.* **1899**, *48*, 19981–19982.

26. Xing, M.; Lv, H.; Ma, J.; Xu, D.; Li, H.; Yang, L.; Kang, J.; Wang, X.; Fang, Z. Transcriptome profiling of resistance to *Fusarium oxysporum* f. sp. *conglutinans* in cabbage (*Brassica oleracea*) roots. *PLoS ONE* **2016**, *11*, e0148048. [CrossRef]
27. Sherf, A.F.; MacNab, A.A. *Vegetable Diseases and Their Control*, 2nd ed.; John Wiley & Sons: New York, NY, USA, 1986; pp. 1–22.
28. Daly, P.; Tomkins, B.; Rural Industries Research and Development Corporation (Canbera, Australia). Production and postharvest handling of Chinese cabbage (*Brassica rapa* var. *pekinensis*): A review of literature. *Rural Ind. Res. Dev. Corp. Barton ACT* **1997**, *97*, 32–35.
29. Pu, Z.; Shimizu, M.; Zhang, Y.; Nagaoka, T.; Hayashi, T.; Hori, H.; Matsumoto, S.; Fujimoto, R.; Okazai, K. Genetic mapping of a Fusarium wilt resistance gene in *Brassica oleracea*. *Mol. Breed.* **2012**, *30*, 809–818. [CrossRef]
30. Shimizu, M.; Pu, Z.; Kawanabe, T.; Kitashiba, H.; Matsumoto, S.; Ebe, Y.; Sano, M.; Funaki, E.; Fujimoto, R.; Okazai, K. Map-based cloning of a candidate gene conferring Fusarium yellows resistance in *Brassica oleracea*. *Theor. Appl. Genet.* **2015**, *128*, 119–130. [CrossRef]
31. Enya, J.; Togawa, M.; Takeuchi, T.; Yoshida, S.; Tsushima, S.; Arie, T.; Sakai, T. Biological and phylogenetic characterization of *Fusarium oxysporum* complex, which causes yellows on *Brassica* spp. and proposal of *F. oxysporum* f. sp. *rapae*, a novel forma specialis pathogenic on *B. rapa* in Japan. *Phytopathology* **2008**, *98*, 475–483. [CrossRef]
32. Lv, H.; Fang, Z.; Yang, L.; Zhang, Y.; Wang, Q.; Liu, Y.; Zhuang, M.; Yang, Y.; Xie, B.; Liu, B.; et al. Mapping and analysis of a novel candidate Fusarium wilt resistance gene *FOC1* in *Brassica oleracea*. *BMC Genom.* **2014**, *15*, 1094. [CrossRef]
33. Chang, A.; Lamara, M.; Wei, Y.; Hu, H.; Parkin, I.A.P.; Gossen, B.D.; Peng, G.; Yu, F. Clubroot resistance gene *Rcr6* in *Brassica nigra* resides in a genomic region homologous to chromosome A08 in *B. rapa*. *BMC Plant Biol.* **2019**, *19*, 224. [CrossRef] [PubMed]
34. Suwabe, K.; Tsukazaki, H.; Iketani, H.; Hatakeyama, K.; Fujimura, M.; Nunome, T.; Fukuoka, H.; Matsumoto, S.; Hirai, M. Identification of two loci for resistance to clubroot (*Plasmodiophora brassicae* Woronin) in *Brassica rapa* L. *Theor. Appl. Genet.* **2003**, *107*, 997–1002. [CrossRef]
35. Sakamoto, K.; Saito, A.; Hayashida, N.; Taguchi, G.; Matsumoto, E. Mapping of isolate-specific QTLs for clubroot resistance in Chinese cabbage (*Brassica rapa* L. spp. *pekinensis*). *Theor. Appl. Genet.* **2008**, *117*, 759–767. [CrossRef] [PubMed]
36. Yu, F.; Zhang, X.; Peng, G.; Falk, K.C.; Strelkov, S.E.; Gossen, B.D. Genotyping-by-sequencing reveals three QTL for clubroot resistance to six pathotypes of *Plasmodiophora brassicae* in *Brassica rapa*. *Sci. Rep.* **2017**, *7*, 4516. [CrossRef]
37. Ueno, H.; Matsumoto, E.; Aruga, D.; Kitagawa, S.; Matumura, H.; Hayashida, N. Molecular characterization of the *CRa* gene conferring clubroot resistance in *Brassica rapa*. *Plant Mol. Biol.* **2012**, *80*, 621–629. [CrossRef]
38. Hatakeyama, K.; Niwa, T.; Kato, T.; Ohara, T.; Kakizaki, T.; Matsumoto, S. The tandem repeated organization of NB-LRR genes in the clubroot-resistant *CRb* locus in *Brassica rapa* L. *Mol. Genet. Genom.* **2017**, *292*, 397–405. [CrossRef]
39. Matsumoto, E.; Yasui, C.; Ohi, M.; Tsukada, M. Linkage analysis of RFLP markers for clubroot resistance and pigmentation in Chinese cabbage (*Brassica rapa* ssp. *pekinensis*). *Euphytica* **1998**, *104*, 79–86. [CrossRef]
40. Hirai, M.; Harada, T.; Kubo, N.; Tsukada, M.; Suwabe, K.; Matsumoto, S. A novel locus for clubroot resistance in *Brassica rapa* and its linkage markers. *Theor. Appl. Genet.* **2004**, *108*, 639–643. [CrossRef]
41. Piao, Z.Y.; Deng, Y.Q.; Choi, S.R.; Park, Y.J.; Lim, Y.P. SCAR and CAPS mapping of CRb, a gene conferring resistance to *Plasmodiophora brassicae* in Chinese cabbage (*Brassica rapa* ssp. *pekinensis*). *Theor. Appl. Genet.* **2004**, *108*, 1458–1465. [CrossRef]
42. Saito, M.; Kubo, N.; Matsumoto, S.; Suwabe, K.; Tsukada, M.; Hirai, M. Fine mapping of the clubroot resistance gene, *Crr3*, in *Brassica rapa*. *Theor. Appl. Genet.* **2006**, *114*, 81–91. [CrossRef]
43. Kato, T.; Hatakeyama, K.; Fukino, N.; Matsumoto, S. Identification of a clubroot resistance locus conferring resistance to a *Plasmodiophora brassicae* classified into pathotype group 3 in Chinese cabbage (*Brassica rapa* L.). *Breed. Sci.* **2012**, *62*, 282–287. [CrossRef] [PubMed]
44. Chen, J.; Jing, J.; Zhan, Z.; Zhang, T.; Zhang, C.; Piao, Z. Identification of novel QTLs for isolate-specific partial resistance to *Plasmodiophora brassicae* in *Brassica rapa*. *PLoS ONE* **2013**, *8*, e85307. [CrossRef] [PubMed]

45. Chu, M.; Song, T.; Falk, K.C.; Zhang, X.; Liu, X.; Chang, A.; Lahlali, R.; McGregor, L.; Gossen, B.D.; Yu, F.; et al. Fine mapping of *Rcr1* and analysis of its effect on transcriptome patterns during infection by *Plasmodiophora brassicae*. BMC Genom. **2014**, *15*, 1166. [CrossRef]
46. Zhang, T.; Zhao, Z.; Zhang, C.; Pang, W.; Choi, S.R.; Lim, Y.P.; Piao, Z. Fine genetic and physical mapping of *CRb* gene conferring resistance to clubroot disease in *Brassica rapa*. Mol. Breed. **2014**, *34*, 1173–1183. [CrossRef]
47. Yu, F.; Zhang, X.; Huang, Z.; Chu, M.; Song, T.; Falk, K.C.; Deora, A.; Chen, Q.; Zhang, Y.; McGregor, L.; et al. Identification of genome-wide variants and discovery of variants associated with *Brassica rapa* clubroot resistance gene *Rcr1* through bulked segregant RNA sequencing. PLoS ONE **2016**, *11*, e0153218. [CrossRef]
48. Huang, Z.; Peng, G.; Liu, X.; Deora, A.; Falk, K.C.; Gossen, B.D.; McDonald, M.R.; Yu, F. Fine mapping of a clubroot resistance gene in Chinese cabbage using SNP markers identified from bulked segregant RNA Sequencing. Front. Plant Sci. **2017**, *8*, 1448. [CrossRef]
49. Pang, W.; Fu, P.; Li, X.; Zhan, Z.; Yu, S.; Piao, Z. Identification and Mapping of the clubroot resistance gene *CRd* in Chinese cabbage (*Brassica rapa* spp. *pekinensis*). Front. Plant Sci. **2018**, *9*, 653. [CrossRef]
50. Kato, T.; Hatakeyama, K.; Fukino, N.; Matsumoto, S. Fine mapping of the clubroot resistance gene *CRb* and development of a useful selectable marker in *Brassica rapa*. Breed. Sci. **2013**, *63*, 116–124. [CrossRef]
51. Nguyen, M.L.; Monakhos, G.F.; Komakhin, R.A.; Monakhos, S.G. The new clubroot resistance locus is located on chromosome A05 in Chinese cabbage (*Brassica rapa* L.). Russ. J. Genet. **2018**, *54*, 296–304. [CrossRef]
52. Suwabe, K.; Tsukazaki, H.; Iketani, H.; Hatakeyama, K.; Kondo, M.; Fujimura, M.; Nunome, T.; Fukuoka, H.; Hirai, M.; Matsumoto, S. Simple sequence repeat-based comparative genomic between *Brassica rapa* and *Arabidopsis thaliana*: The genetic origin of clubroot resistance. Genetics **2006**, *173*, 309–319. [CrossRef]
53. Hatakeyama, K.; Suwabe, K.; Tomita, R.N.; Kato, T.; Nunome, T.; Fukuoka, H.; Matsumoto, S. Identification and characterization of *Crr1a*, a gene for resistance to clubroot disease (*Plasmodiophora brassicae* Woronin) in *Brassica rapa* L. PLoS ONE **2013**, *8*, e54745. [CrossRef] [PubMed]
54. Laila, R.; Park, J.I.; Robin, A.H.K.; Natarajan, S.; Vijayakumar, H.; Shirasawa, K.; Isobe, S.; Kim, H.T.; Nou, I.S. Mapping of a novel clubroot resistance QTL using ddRAD-seq in Chinese cabbage (*Brassica rapa* L.). BMC Plant Biol. **2019**, *19*, 13. [CrossRef]
55. Kuginuki, Y.; Ajisaka, H.; Yui, M.; Yoshikawa, H.; Hida, K.; Hirai, M. RAPD markers linked to a clubroot-resistance locus in *Brassica rapa* L. Euphytica **1997**, *98*, 149–154. [CrossRef]
56. Lan, M.; Li, G.; Hu, J.; Yang, H.; Zhang, L.; Xu, X.; Liu, J.; He, J.; Sun, R. iTRAQ-based quantitative analysis reveals proteomic changes in Chinese cabbage (*Brassica rapa* L.) response to *Plasmodiophora brassicae* infection. Sci. Rep. **2019**, *9*, 12058. [CrossRef]
57. Dakouri, A.; Zhang, X.; Peng, G.; Falk, K.C.; Gossen, B.D.; Strelkov, S.E.; Yu, F. Analysis of genome-wide variants through bulked segregant RNA sequencing reveals a major gene of resistance to *Plasmodiophora brassicae* in *Brassica oleracea*. Sci. Rep. **2018**, *8*, 17657. [CrossRef]
58. Figdore, S.S.; Ferreia, M.E.; Slocum, M.K.; Williams, P.H. Association of RFLP markers with trait loci affecting clubroot resistance and morphological characters in *Brassica oleracea* L. Euphytica **1993**, *69*, 33–44. [CrossRef]
59. Grandclement, C.; Laurens, F.; Thomas, G. Genetic analysis of resistance to clubroot (*Plasmodiophora brassicae* Woron) in two *Brassica oleracea* groups (spp. *acephala* and spp. *botrytis*) through diallel analysis. Plant Breed. **1996**, *115*, 152–156. [CrossRef]
60. Voorrips, R.E.; Jongerius, M.C.; Kanne, H.J. Mapping of two genes for resistance to clubroot (*Plasmodiophora brassicae*) in a population of double haploid lines of *Brassica oleracea* by means of RFLP and AFLP markers. Theor. Appl. Genet. **1997**, *94*, 75–82. [CrossRef]
61. Moriguchi, K.; Kimizuka-Takagi, C.; Ishii, K.; Nomura, K. A genetic map based on RAPD, RFLP, isozyme, morphological markers and QTL analysis for clubroot resistance in *Brassica oleracea*. Breed. Sci. **1999**, *49*, 257–265. [CrossRef]
62. Nomura, K.; Minegishi, Y.; Kimizuka-Takagi, C.; Fujioka, T.; Moriguchi, K.; Shishido, R.; Ikehashi, H. Evaluation of F_2 and F_3 plants introgressed with QTLs for clubroot resistance in cabbage developed by using SCAR markers. Plant Breed. **2005**, *124*, 371–375. [CrossRef]
63. Rocherieux, J.; Glory, P.; Giboulot, A.; Boury, S.; Barbeyron, G.; Thomas, G.; Manzanares-Dauleux, M.J. Isolate-specific and broad-spectrum QTLs are involved in the control of clubroot in *Brassica oleracea*. Theor. Appl. Genet. **2004**, *108*, 1555–1563. [CrossRef]

64. Nagaoka, T.; Doullah, M.A.U.; Matsumoto, S.; Kawasaki, S.; Ishikawa, T.; Hori, H.; Okazaki, K. Identification of QTLs that control clubroot resistance in *Brassica oleracea* and comparative analysis of clubroot resistance genes between *B. rapa* and *B. oleracea*. *Theor. Appl. Genet.* **2010**, *120*, 1335–1346. [CrossRef] [PubMed]
65. Lee, J.; Izzah, N.K.; Choi, B.S.; Joh, H.J.; Lee, S.C.; Perumal, S.; Seo, J.; Ahn, K.; Jo, E.J.; Choi, G.J.; et al. Genotyping-by-sequencing map permits identification of clubroot resistance QTLs and revision of the reference genome assembly in cabbage (*Brassica oleracea* L.). *DNA Res.* **2016**, *23*, 29–41. [CrossRef] [PubMed]
66. Peng, L.; Zhou, L.; Li, Q.; Wei, D.; Ren, X.; Song, H.; Mei, J.; Si, J.; Qian, W. Identification of quantitative trait loci for clubroot resistance in *Brassica oleracea* with the use of *Brassica* SNP microarray. *Front. Plant Sci.* **2018**, *9*, 822. [CrossRef] [PubMed]
67. Tomita, H.; Shimizu, M.; Doullah, M.A.U.; Fujimoto, R.; Okazaki, K. Accumulation of quantitative trait loci conferring broad-spectrum clubroot resistance in *Brassica oleracea*. *Mol. Breed.* **2013**, *32*, 889–900. [CrossRef]
68. Landry, B.S.; Hubert, N.; Crete, R.; Chang, M.S.; Lincoln, S.E.; Etoh, T. A genetic map for *Brassica oleracea* based on RFLP markers detected with expressed DNA sequences and mapping of resistance genes to race 2 of *Plasmodiophora brassicae* (Woronin). *Genome* **1992**, *35*, 409–420. [CrossRef]
69. Manzanares-Dauleux, M.J.; Delourme, R.; Baron, F.; Tomas, G. Mapping of one major gene and of QTLs involved in resistance to clubroot in *Brassica napus*. *Theor. Appl. Genet.* **2000**, *101*, 885–891. [CrossRef]
70. Werner, S.; Diederichsen, E.; Frauen, M.; Schondelmaier, J.; Jung, C. Genetic mapping of clubroot resistance genes in oilseed rape. *Theor. Appl. Genet.* **2008**, *116*, 363–372. [CrossRef]
71. Fredua-Agyeman, R.; Rahman, R. Mapping of the clubroot disease resistance in spring *Brassica napus* canola introgressed from European winter canola cv. 'Mendel'. *Euphytica* **2016**, *211*, 201–213. [CrossRef]
72. Diederichsen, E.; Beckmann, J.; Schondelmeier, J.; Dreyer, F. Genetics of clubroot resistance in *Brassica napus* 'Mendel'. *Acta. Hort.* **2006**, *706*, 307–312. [CrossRef]
73. Zhang, X.; Feng, J.; Hwang, S.F.; Strelkov, S.E.; Falak, I.; Huang, X.; Sun, R. Mapping of clubroot (*Plasmodiophora brassicae*) resistance in canola (*Brassica napus*). *Plant Pathol.* **2016**, *65*, 435–440. [CrossRef]
74. Hasan, M.J.; Rahman, R. Genetics and molecular mapping of resistance to *Plasmodiophora brassicae* pathotypes 2, 3, 5, 6 and 8 in rutabaga (*Brassica napus* var. *napobrassica*). *Genome* **2016**, *59*, 805–815. [CrossRef]
75. Hejna, O.; Havlickova, L.; He, Z.; Bancroft, I.; Curn, V. Analysing the genetic architecture of clubroot resistance variation in *Brassica napus* by associative transcriptomics. *Mol. Breed.* **2019**, *39*, 112. [CrossRef] [PubMed]
76. Blank, L.M. Fusarium resistance in Wisconsin all seasons cabbage. *J. Agric. Res.* **1937**, *55*, 497–510.
77. Walker, J.C. *Plant Disease-Vegetable Crops-Cauliflower, Cabbage, and Others*; US Department of Agriculture, US Government Printing Office: Washington, DC, USA, 1953.
78. Lv, H.; Yang, L.; Kang, J.; Wang, Q.; Wang, X.; Fang, Z.; Liu, Y.; Zhuang, M.; Zhang, Y.; Lin, Y.; et al. Development of InDel markers linked to Fusarium wilt resistance in cabbage. *Mol. Breed.* **2013**, *32*, 961–967. [CrossRef]
79. Miyaji, N.; Shimizu, M.; Miyazaki, J.; Osabe, K.; Sato, M.; Ebe, Y.; Takada, S.; Kaji, M.; Dennis, E.S.; Fujimoto, R.; et al. Comparison of transcriptome profiles by *Fusarium oxysporum* inoculation between Fusarium yellows resistant and susceptible lines in *Brassica rapa* L. *Plant Cell Rep.* **2017**, *36*, 1841–1854. [CrossRef] [PubMed]
80. Sato, M.; Shimizu, M.; Shea, D.J.; Hoque, M.; Kawanabe, T.; Miyaji, N.; Fujimoto, R.; Fukai, E.; Okazaki, K. Allele specific DNA marker for fusarium resistance gene *FocBo1* in *Brassica oleracea*. *Breed. Sci.* **2019**, *69*, 308–315. [CrossRef]
81. Shimizu, M.; Fujimoto, R.; Ying, H.; Pu, Z.; Ebe, Y.; Kawanabe, T.; Saeki, N.; Taylor, J.M.; Kaji, M.; Dennis, E.S.; et al. Identification of candidate genes for Fusarium yellows resistance in Chinese cabbage by differential expression analysis. *Plant Mol. Biol.* **2014**, *85*, 247–257. [CrossRef] [PubMed]
82. Kawamura, K.; Kawanabe, T.; Shimizu, M.; Nagano, A.J.; Saeki, N.; Okazaki, K.; Kaji, M.; Dennis, E.S.; Osabe, K.; Fujimoto, R. Genetic characterization of inbred lines of Chinese cabbage by DNA markers; towards the application of DNA markers to breeding of F_1 hybrid cultivars. *Data Brief* **2016**, *6*, 229–237. [CrossRef] [PubMed]
83. Pu, Z.; Ino, Y.; Kimura, Y.; Tago, A.; Shimizu, M.; Natsume, S.; Sano, Y.; Fujjimoto, R.; Kaneko, K.; Shea, D.J.; et al. Changes in the proteome of xylem sap in *Brassica oleracea* in response to *Fusarium oxysporum* stress. *Front. Plant Sci.* **2016**, *7*, 31. [CrossRef] [PubMed]

84. Kawamura, K.; Shimizu, M.; Kawanabe, T.; Pu, Z.; Kodama, T.; Kaji, M.; Kenji, O.; Fujimoto, R.; Keiichi, O. Assessment of DNA markers for seed contamination testing and selection of disease resistance in cabbage. *Euphytica* **2017**, *213*, 28. [CrossRef]
85. Matsumoto, E.; Ueno, H.; Aruga, D.; Sakamoto, K.; Hayashida, N. Accumulation of three clubroot resistance genes through marker-assisted selection in Chinese cabbage (*Brassica rapa* spp. *pekinensis*). *J. Jpn. Soc. Hortic. Sci.* **2012**, *81*, 184–190. [CrossRef]
86. Doullah, M.A.U.; Tomita, H.; Shimizu, M.; Matsumoto, S.; Fujimoto, R.; Okazaki, K. Recent progress of clubroot resistance breeding through marker assisted selection in *Brassica rapa* and *Brassica oleracea*. *J. Sylhet Agric. Univ.* **2014**, *1*, 139–146.
87. Strelkov, S.E.; Hwang, S.F.; Manolii, V.P.; Cao, T.; Fredua-Agyeman, R.; Harding, M.W.; Peng, G.; Gossen, B.D.; Mcdonald, M.R.; Feindel, D. Virulence and pathotype classification of *Plasmodiophora brassicae* populations collected from clubroot resistant canola (*Brassica napus*) in Canada. *Can. J. Plant Pathol.* **2018**, *40*, 284–298. [CrossRef]
88. Kuginuki, Y.; Yoshikawa, H.; Hirai, M. Variation in virulence of *Plasmodiophora brassicae* in Japan tested with clubroot-resistant cultivars of Chinese cabbage (*Brassica rapa* L. spp. *pekinensis*). *Eur. J. Plant Pathol.* **1999**, *105*, 327–332. [CrossRef]
89. Kou, Y.; Wang, S. Broad-spectrum and durability: Understanding of quantitative disease resistance. *Curr. Opin. Plant Biol.* **2010**, *13*, 181–185. [CrossRef]
90. Crisp, P.; Crute, I.R.; Sutherland, R.A.; Angell, S.M.; Bloor, K.; Burgess, H.; Gordon, P.L. The exploitation of genetic resources of *Brassica oleracea* in breeding for resistance to clubroot (*Plasmodiophora brassicae*). *Euphytica* **1889**, *42*, 215–226.
91. Shah, N.; Sun, J.; Yu, S.; Yang, Z.; Wang, Z.; Huang, F.; Dun, B.; Gong, J.; Liu, Y.; Li, Y.; et al. Genetic variation analysis of field isolates of clubroot and their responses to *Brassica napus* lines containing resistant genes *CRb* and *PbBa8.1* and their combination in homozygous and heterozygous state. *Mol. Breed.* **2019**, *39*, 153. [CrossRef]
92. Rahman, H.; Franke, C. Association of fusarium wilt susceptibility with clubroot resistance derived from winter *Brassica napus* L. 'Mendel'. *Can. J. Plant Pathol.* **2019**, *41*, 60–64. [CrossRef]

© 2020 by the authors. Licensee MDPI, Basel, Switzerland. This article is an open access article distributed under the terms and conditions of the Creative Commons Attribution (CC BY) license (http://creativecommons.org/licenses/by/4.0/).

Article

Development of a New DNA Marker for Fusarium Yellows Resistance in *Brassica rapa* Vegetables

Naomi Miyaji [1,2,*,†], Mst Arjina Akter [1,3,†], Chizuko Suzukamo [4], Hasan Mehraj [1], Tomoe Shindo [5], Takeru Itabashi [5], Keiichi Okazaki [6], Motoki Shimizu [2], Makoto Kaji [4], Masahiko Katsumata [4], Elizabeth S. Dennis [7,8] and Ryo Fujimoto [1,*]

1. Graduate School of Agricultural Science, Kobe University, Rokkodai, Nada-ku, Kobe 657-8501, Japan; 189a362a@stu.kobe-u.ac.jp (M.A.A.); hmehraj34@stu.kobe-u.ac.jp (H.M.)
2. Iwate Biotechnology Research Center, Narita, Kitakami, Iwate 024-0003, Japan; m-shimizu@ibrc.or.jp
3. Department of Plant Pathology, Bangladesh Agricultural University, Mymensingh 2202, Bangladesh
4. Watanabe seed Co., Ltd., Machiyashiki, Misato-cho, Miyagi 987-0003, Japan; ws-ken@chive.ocn.ne.jp (C.S.); ws-kaji@chive.ocn.ne.jp (M.K.); ws-katsu@aroma.ocn.ne.jp (M.K.)
5. Miyagi Prefectural Agriculture and Horticulture Research Center, Natori, Miyagi 981-1243, Japan; shindo-to469@pref.miyagi.lg.jp (T.S.); itabashi-ta714@pref.miyagi.lg.jp (T.I.)
6. Graduate School of Science and Technology, Niigata University, Ikarashi, Nishi-ku, Niigata 950-2181, Japan; okazaki@agr.niigata-u.ac.jp
7. CSIRO Agriculture and Food, Canberra, ACT 2601, Australia; Liz.Dennis@csiro.au
8. School of Life Science, Faculty of Science, University of Technology Sydney, P.O. Box 123, Broadway, NSW 2007, Australia
* Correspondence: n-miyaji@ibrc.or.jp (N.M.); leo@people.kobe-u.ac.jp (R.F.)
† Authors contributed equal.

Citation: Miyaji, N.; Akter, M.A.; Suzukamo, C.; Mehraj, H.; Shindo, T.; Itabashi, T.; Okazaki, K.; Shimizu, M.; Kaji, M.; Katsumata, M.; et al. Development of a New DNA Marker for Fusarium Yellows Resistance in *Brassica rapa* Vegetables. *Plants* **2021**, *10*, 1082. https://doi.org/10.3390/plants10061082

Academic Editor: Srinivasan Ramachandran

Received: 30 April 2021
Accepted: 24 May 2021
Published: 27 May 2021

Publisher's Note: MDPI stays neutral with regard to jurisdictional claims in published maps and institutional affiliations.

Copyright: © 2021 by the authors. Licensee MDPI, Basel, Switzerland. This article is an open access article distributed under the terms and conditions of the Creative Commons Attribution (CC BY) license (https:// creativecommons.org/licenses/by/ 4.0/).

Abstract: In vegetables of *Brassica rapa* L., *Fusarium oxysporum* f. sp. *rapae* (*For*) or *F. oxysporum* f. sp. *conglutinans* (*Foc*) cause Fusarium yellows. A resistance gene against *Foc* (*FocBr1*) has been identified, and deletion of this gene results in susceptibility (*focbr1-1*). In contrast, a resistance gene against *For* has not been identified. Inoculation tests showed that lines resistant to *Foc* were also resistant to *For*, and lines susceptible to *Foc* were susceptible to *For*. However, prediction of disease resistance by a dominant DNA marker on *FocBr1* (Bra012688m) was not associated with disease resistance of *For* in some komatsuna lines using an inoculation test. QTL-seq using four F$_2$ populations derived from *For* susceptible and resistant lines showed one causative locus on chromosome A03, which covers *FocBr1*. Comparison of the amino acid sequence of *FocBr1* between susceptible and resistant alleles (*FocBr1* and *FocBo1*) showed that six amino acid differences were specific to susceptible lines. The presence and absence of *FocBr1* is consistent with *For* resistance in F$_2$ populations. These results indicate that *FocBr1* is essential for *For* resistance, and changed amino acid sequences result in susceptibility to *For*. This susceptible allele is termed *focbr1-2*, and a new DNA marker (focbr1-2m) for detection of the *focbr1-2* allele was developed.

Keywords: Fusarium yellows; *Fusarium oxysporum* f. sp. *rapae*; DNA marker; *R* gene; marker-assisted selection; QTL-seq; *Brassica rapa*

1. Introduction

Brassica rapa L. comprises a variety of vegetables that are rich sources of nutrients including vitamins, minerals, dietary fiber, and phytochemicals [1,2]. In leafy vegetables of *B. rapa*, there are two morphotypes, heading types such as Chinese cabbage (var. *pekinensis*) and non-heading type such as pak choi (var. *chinensis*), komatsuna (var. *perviridis*) or chijimina (var. *narinosa*). Root vegetables such as turnip (var. *rapa*) also belong to *B. rapa* [1,2]. Most commercial cultivars of these vegetables are F$_1$ hybrids, and hybrid vigor, disease resistance, and late bolting are important breeding traits [3–5]. In particular, disease resistance is demanded by farmers, especially for soil-borne diseases that are difficult to control with chemicals [5,6].

Plants have evolved their immunity to pathogens via two mechanisms [7,8]. Pattern recognition receptors (PRRs) located in the plant cell membrane recognize pathogen-associated molecular patterns (PAMPs), which activate PAMP-triggered immunity (PTI) and restrict pathogen development. Most pathogens secrete effectors (avirulence (AVR) proteins) into plant cells to suppress PTI, while plants have various *resistance* (*R*) genes, which mainly encode Toll/Interleukin-1 receptor (TIR) or coiled-coil (CC), nucleotide-binding site (NBS), and leucine-rich repeat (LRR) domains, to detect effectors. Recognition of effectors by R proteins induces effector-triggered immunity (ETI), and recognition of specific effectors by R proteins is termed "gene-for-gene resistance" [7].

Fusarium oxysporum is a soil-borne fungus and comprises 150 host-specific formae speciales. *F. oxysporum* causes yellows in a wide range of host plants [5,6]. In *B. rapa* vegetables, two formae speciales of *F. oxysporum* f. sp. *conglutinans* (*Foc*) and f. sp. *rapae* (*For*) have been identified as causing Fusarium yellows [9]. *Foc* was first reported as a causal agent of yellowing in cabbage (*Brassica oleracea* L. var. *capitata*) in 1913 [5,6] and causes Fusarium yellows not only in *B. oleracea* vegetables including cabbage or broccoli but also in *B. rapa* vegetables including turnip, komatsuna, and pak choi [9–13]. In contrast, *For* causes yellowing in *B. rapa* vegetables, but not in *B. oleracea* vegetables [9].

The Fusarium yellows *R* gene against *Foc* has been identified in *B. rapa* (*FocBr1*) and *B. oleracea* (*FocBo1*) [11–14]. *FocBr1* and *FocBo1* are orthologs and encode a TIR-NBS-LRR protein. In *B. rapa*, an approximately 35-kb deletion including *FocBr1* results in susceptibility (*focbr1-1*) and there are no reports of other causative mutations for susceptibility [12,15]. In contrast, there are three different susceptible alleles of *FocBo1* in *B. oleracea* (*focbo1-1*, *focbo1-2* and *focbo1-3*), but a 35-kb deletion similar to that in Fusarium yellows susceptible lines of *B. rapa* has not been identified [11,13,16,17].

The Fusarium yellows *R* gene against *For* (*ForBr1*) has not been identified. In this study, we performed QTL-seq to isolate *ForBr1*. We developed a DNA marker that can identify the susceptible alleles of Fusarium yellows and tested this marker in cultivars of *B. rapa* vegetables.

2. Results

2.1. Screening of Lines for Resistance to F. oxysporum f. sp. Rapae

We have shown that *FocBr1* (Bra012688) is a resistance gene to *F. oxysporum* f. sp. *conglutinans* (*Foc*), and deletion of this gene results in susceptibility to *Foc* [12]; this susceptible allele is termed *focbr1-1* [14]. We made a DNA marker (Bra012688m) to detect the deletion of *FocBr1* that is homozygous for the *focbr1-1* allele [12,15]. In this study, we performed inoculation tests using *F. oxysporum* f. sp. *rapae* (*For*) for screening for resistant lines. Three Chinese cabbage (var. *pekinensis*), three turnip (var. *rapa*) and 22 komatsuna (var. *perviridis*) lines were tested of which 18 lines were resistant and 10 lines were susceptible (Table 1). We also inoculated these 28 lines with *Foc*, and resistance to *For* and *Foc* was identical (Table 1). We examined whether the results of the inoculation test were consistent with the prediction by *FocBr1* DNA marker (Bra012688m). In all lines of Chinese cabbage and turnip, the prediction by the DNA marker was identical to the resistance determined by the inoculation test, while in seven of 22 komatsuna lines ("Zaoh", YBCG-12, YBCG-13, YBCG-14, YBCG-15, YBCG-TC02, and YBCG-TC05) the DNA marker prediction was not consistent with the results of the inoculation test (Table 1). We tested an additional 15 lines of *B. rapa*; three lines ("Chijimikomatsuna", "Tsunashima", and "Hirose") were not consistent between the DNA marker prediction and the results of the inoculation test using *For* (Table S1).

Table 1. Assessment of Fusarium yellows resistance by inoculation test.

Name	Inoculation Test		Prediction by DNA Marker
	For	Foc	Bra012688m
Chinese cabbage (var. *pekinensis*)			
"W77"	R	R	+
RJKB-T23	R	R	+
RJKB-T24	S	S	-
Turnip (var. *rapa*)			
"CR-Yukiakari"	R *	R *	+
"Hekiju"	R	R	+
NSI-01	R *	R *	+
Komatsuna (var. *perviridis*)			
"CR-Taiga"	R *	R *	+
"Manaka"	R	R	+
"Nanami"	R	R	+
"Natsurakuten"	R *	R *	+
"Zaoh"	S	S	+
YBCG-08	R	R	+
YBCG-09	S	S	-
YBCG-10	S	S	-
YBCG-11	R	R	+
YBCG-12	S	S	+
YBCG-13	S	S	+
YBCG-14	S	S	+
YBCG-15	S	S	+
YBCG-16	R	R	+
YBCG-17	R	R	+
YBCG-18	R *	R *	+
YBCG-TC01	R	R	+
YBCG-TC02	S	S	+
YBCG-TC03	R	R	+
YBCG-TC04	R *	R *	+
YBCG-TC05	S	S	+
YBCG-TC06	R	R	+

R and S represent resistant and susceptible, respectively, to *Foc* or *For*. * represents weak resistance (some of the 25 seedlings showed IP = 0, while others showed IP ≥ 3). +, amplification by Bra012688m; -, no amplification by Bra012688m.

2.2. Identification of the Causative Region of Resistance for F. oxysporum f. sp. Rapae

We performed linkage analysis using three individual F_2 populations derived from hybrids between *For* susceptible lines not containing the *FocBr1* deletion and resistant lines. In the 200 plants of the F_2 population derived from YBCG-11 (resistant) × YBCG-12 (susceptible) hybrid, 169 plants were resistant and 31 plants were susceptible to *For*. The number of susceptible plants was too small to be explained by a single gene dominance (chi-squared test, $p < 0.05$) (Table 2). This was also the case for the other two populations derived from YBCG-11 × YBCG-13 (susceptible) and YBCG-11 × YBCG-14 (susceptible) hybrids (Table 2). To identify the region covering the *R* gene for *For* (*ForBr1*), we performed QTL-seq analysis using bulked DNAs derived from about 20 resistant and susceptible individual plants derived from YBCG-11 × YBCG-12, YBCG-11 × YBCG-13, or YBCG-11 × YBCG-14 hybrids and found one similar locus on chromosome A03 in all three populations (Figure 1, Figures S1–S3). 22.0–33.5 Mb, 22.9–35.5 Mb, and 22.5–33.6 Mb region was detected as the QTL by 95% significance in the F_2 population derived from YBCG-11 × YBCG-12, YBCG-11 × YBCG-13, and YBCG-11 × YBCG-14 hybrid, respectively (Figure 1). 1824, 1778, and 1734 genes were located in three QTLs, and 1655 genes overlapped (Figure S4, Table S2). A domain search using HMMSCAN with Pfam database and NCBI conserved domain search found nine genes encoding NBS-LRR proteins, including *FocBr1* (BraA03g047240.3C or Bra012688) (Table S3).

Table 2. Linkage analysis using six individual F$_2$ populations.

	F$_2$ Population		χ^2 (R:S = 3:1)	Resistant Parent	Bra012688m	Susceptible Parent	Bra012688m
	Resistant	Susceptible					
1	169	31	$p < 0.05$	YBCG-11	+	YBCG-12	+
2	171	29	$p < 0.001$	YBCG-11	+	YBCG-13	+
3	171	29	$p < 0.001$	YBCG-11	+	YBCG-14	+
4	160	40	$p > 0.05$	YBCG-08	+	YBCG-09	-
5	149	51	$p > 0.05$	YBCG-TC01	+	YBCG-10	-
6	156	44	$p > 0.05$	YBCG-11	+	YBCG-10	-

+; amplification by Bra012688m, -; no amplification by Bra012688m.

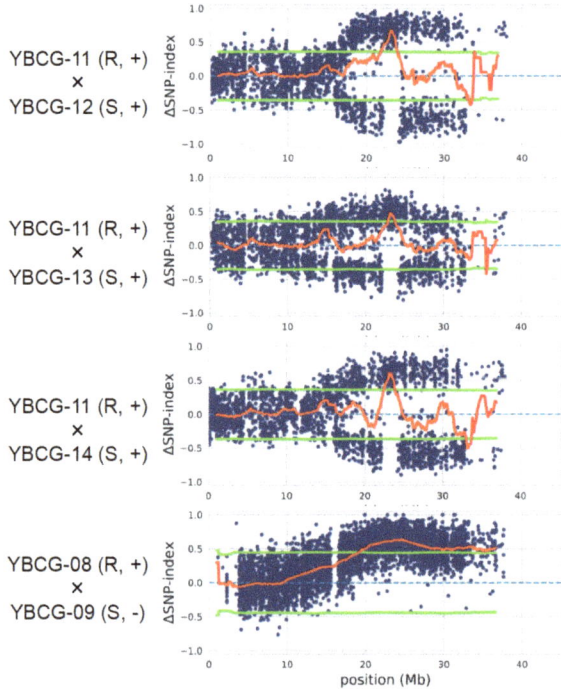

Figure 1. QTL-seq results on chromosome A03. F$_2$ populations derived from YBCG-11 × YBCG-12, YBCG-11 × YBCG-13, YBCG-11 × YBCG-14, and YBCG-08 × YBCG-09 hybrids were used. Blue dots indicate ΔSNP-index, and the red line indicates the sliding window average of ΔSNP-index. Light green lines represent $p < 0.05$. R and S represent resistant and susceptible, respectively. + and - represent the presence and absence of PCR amplification of Bra012688m marker, respectively.

2.3. A New Susceptible Allele of FocBr1 Was Identified

Because *FocBr1* was included in three QTLs, we focused on *FocBr1* for further analysis. The expression level of *FocBr1* in three susceptible lines ("Zaoh", YBCG-12, and YBCG-15) was similar to that of resistant lines (YBCG-11 and YBCG-16) (Figure 2), indicating that expression levels are not related to susceptibility. Next, we compared the amino acid sequences of FocBr1 in resistant and susceptible lines. The amino acid sequence of FocBr1 in the resistant line, YBCG-11, was 100% identical to FocBr1 in the resistant line, RJKB-T23 [12]. Amino acid sequences of FocBr1 were 100% identical among bulked DNAs of susceptible plants derived from F$_2$ populations of YBCG-11 × YBCG-12, YBCG-11 × YBCG-13 and YBCG-11 × YBCG-14 hybrids, but there were some substitutions of amino acid sequences compared with FocBr1 in YBCG-11 (Figure 3). There were eleven amino acid sequence differences in FocBr1 between resistant and susceptible lines; five (A546T, N721D, T803K, V805E, and K862N) of which were identical between FocBr1 in the susceptible lines and the resistant allele of FocBo1 (*B. oleracea*) (Figure 3). Both FocBr1 and FocBo1 are resistance genes to *Foc*, indicating that the difference of amino acid sequences between FocBr1 and FocBo1 might not relate to the Fusarium yellows resistance, and the identical amino acid sequences between susceptible lines and FocBo1 might not lead to its susceptibility. The remaining six amino acid changes (Q859W, M869K, L1060F, V1148L, K1212T, and Q1395L) were differed between susceptible lines (YBCG-12, YBCG-13, and YBCG-14) and resistance lines of *B. rapa* and *B. oleracea* (Figure 3), which are susceptible line specific, suggesting that some of these amino acid sequences specific to susceptible lines result in susceptibility to *For*; some mutations may result in loss of function.

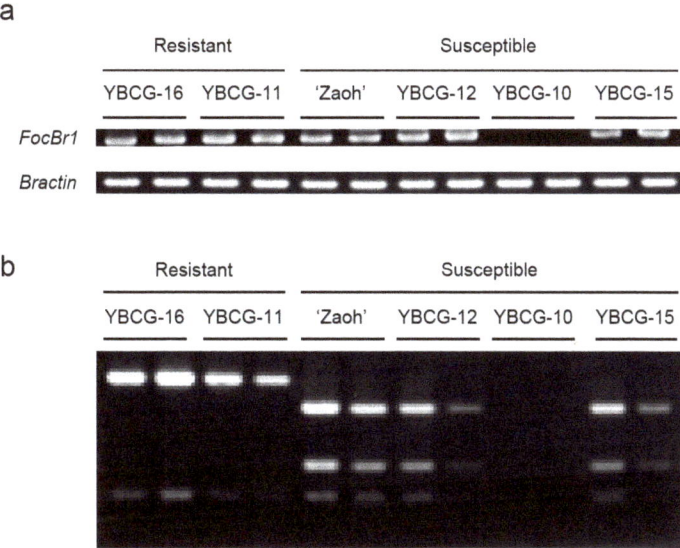

Figure 2. Expression and genotype of *FocBr1* in *For* resistant and susceptible lines. (**a**) Expression of *FocBr1* and *Bractin* (control) was confirmed by RT-PCR. (**b**) DNA fragments of RT-PCR products digested by *Hin*d III. YBCG-16 and YBCG-11 have *FocBr1/FocBr1* homozygous or *FocBr1/focbr1-1* heterozygous alleles. "Zaoh", YBCG-12 and YBCG-15 have *focbr1-2/focbr1-2* homozygous or *focbr1-2/focbr1-1* heterozygous alleles, and YBCG-10 has *focbr1-1/focbr1-1* homozygous allele.

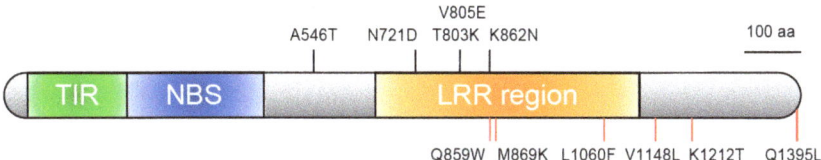

Figure 3. Protein structure of FocBr1 in the resistant line of *B. rapa*. TIR (green box), NBS (blue box), and LRR region (orange box) were identified. Black lines represent the position of difference of amino acid sequences between resistant and susceptible lines in *B. rapa*, while amino acid sequences of FocBr1 in the susceptible lines were identical to the FocBo1 (*Foc* resistance gene in *B. oleracea*). Red lines represent the position of susceptible line-specific amino acid substitutions. Domains were predicted using HMMSCAN with Pfam database. (https://www.ebi.ac.uk/Tools/hmmer/, accessed on 1 April 2021) and NCBI conserved domain search (https://www.ncbi.nlm.nih.gov/Structure/cdd/wrpsb.cgi, accessed on 1 April 2021).

We examined whether *FocBr1* deletion (*focbr1-1*) causes susceptibility to *For*. Linkage analysis using three individual F_2 populations derived from hybrids between *For* susceptible lines (*focbr1-1*) and resistant lines (YBCG-08 (resistant) × YBCG-09 (susceptible), YBCG-TC01 (resistant) × YBCG-10 (susceptible), and YBCG-11 × YBCG-10) showed that the number of resistant and susceptible plants segregated as 3:1 ratio (chi-squared test, $p > 0.05$) (Table 2). QTL-seq analysis using bulked DNAs derived from about 20 resistant and susceptible plants of the F_2 population derived from YBCG-08 × YBCG-09 hybrid found one causative locus (19.0–36.6 Mb) on chromosome A03, which covers the *FocBr1* locus (Figure 1 and Figure S5). We tested a DNA marker (Bra012688m) in 12 resistant and susceptible plants from these three F_2 populations, and the presence and absence of *FocBr1* was consistent with the inoculation test (Figure S6). These results indicate that *FocBr1* is essential for resistance to not only *Foc* but also *For*, supporting the suggestion that mutations cause susceptibility to *Foc* and *For*. This susceptible allele was termed *focbr1-2*.

2.4. Development of a New DNA Marker for Fusarium Yellows Resistance

Using sequence polymorphism between *FocBr1* and *focbr1-2*, a new cleaved amplified polymorphic sequence (CAPS) DNA marker (focbr1-2m) was developed. Using this DNA marker, genotypes of 12 resistant and susceptible plants in three F_2 populations derived from YBCG-11 × YBCG-12, YBCG-11 × YBCG-13, and YBCG-11 × YBCG-14 hybrids were confirmed, and genotypes were identical to the resistance determined by the inoculation test (Table S4). All ten lines that were not consistent between *For* inoculation test and the DNA marker (Bra012688m) prediction had homozygous *focbr1-2* or heterozygous *focbr1-2* and *focbr1-1* alleles (Tables 1 and 3, and Table S1). Another 33 *B. rapa* lines were consistent between the inoculation test using *For* and prediction by DNA marker (focbr1-2m) (Table 3). These results indicate that this new DNA marker can detect not only the *focbr1-1* susceptible allele but also the *focbr1-2* allele.

2.5. Prediction of Fusarium Yellows Resistance in Commercial B. rapa Vegetables by DNA Marker

Using the focbr1-2m marker, we predicted the resistance to *For* in 157 cultivars of Chinese cabbage, 35 cultivars of turnip, 40 cultivars of pak choi, and 73 cultivars of komatsuna. Of 157 cultivars of Chinese cabbage, six cultivars (3.8%) were heterozygous for *FocBr1*/*focbr1-2*, and there were no cultivars homozygous for *focbr1-2*/*focbr1-2* or heterozygous for *focbr1-2*/*focbr1-1* (Table 4). There were six Chinese cabbage cultivars (3.8%) homozygous for *focbr1-1*/*focbr1-1* (Table 4), which could be susceptible to either *For* or *Foc*. Of 35 cultivars of turnip, five cultivars (14.3 %) were heterozygous for *FocBr1*/*focbr1-2*, and there were no cultivars homozygous for *focbr1-2*/*focbr1-2*, *focbr1-1*/*focbr1-1*, or heterozygous for *focbr1-2*/*focbr1-1* (Table 4). Of 40 cultivars of pak choi, 16 cultivars (40.0%) were heterozygous for *FocBr1*/*focbr1-2*, and there were no cultivars homozygous for *focbr1-2*/*focbr1-2*, *focbr1-1*/*focbr1-1*, or heterozygous for *focbr1-2*/*focbr1-1* (Table 4). Of 73 cultivars of komatsuna, 21 cultivars (28.8%) were heterozygous for *FocBr1*/*focbr1-2*, and five cultivars (6.8%) were homozygous for *focbr1-2*/*focbr1-2* or heterozygous for *focbr1-2*/*focbr1-1*. There were three cultivars (4.1%) homozygous for *focbr1-1*/*focbr1-1* (Table 4). Cultivars with *focbr1-2*/*focbr1-2* or *focbr1-2*/*focbr1-1* were found only in komatsuna.

Table 3. *F. oxysporum* f. sp. *rapae* resistance and *FocBr1* genotype determined by focbr1-2m marker.

Name	Inoculation Test	Prediction by DNA Markers	
	For	Bra012688m	focbr1-2m
Chinese cabbage (var. *pekinensis*)			
"W77"	R	+	A
RJKB-T23	R	+	A
RJKB-T24	S	-	D
RJKB-T36	R	+	A
RJKB-T37	R	+	A
RJKB-T38	R	+	A
RJKB-T39	R	+	A
RJKB-T40	S	-	D
Turnip (var. *rapa*)			
"CR-Yukiakari"	R*	+	A
"Hekiju"	R	+	A
"Yukibotan"	R	+	A
NSI-01	R*	+	A
Pak choi (var. *chinensis*)			
"Entei"	R	+	C
"Ryoutou"	R	+	A
Komatsuna (var. *perviridis*)			
"Chijimikomatsuna"	S	+	B
"CR-Taiga"	R*	+	C
"Kahoku"	R	+	C
"Manaka"	R	+	C
"Nakamachi"	R	+	A
"Nanami"	R	+	A
"Nanane"	R	+	A
"Natsurakuten"	R*	+	C
"Norichan"	R	+	A
"Tsunashima"	S	+	B
"Zaoh"	S	+	B
YBCG-08	R	+	A
YBCG-09	S	-	D
YBCG-10	S	-	D
YBCG-11	R	+	A
YBCG-12	S	+	B
YBCG-13	S	+	B
YBCG-14	S	+	B
YBCG-15	S	+	B
YBCG-16	R	+	A
YBCG-17	R	+	A
YBCG-18	R*	+	A
YBCG-TC01	R	+	A
YBCG-TC02	S	+	B
YBCG-TC03	R	+	A
YBCG-TC04	R*	+	C
YBCG-TC05	S	+	B
YBCG-TC06	R	+	A
Chijimina (var. *narinosa*)			
"Hirose"	S	+	B

R and S represent resistant and susceptible to *For*, respectively. * represents weak resistance (some of the 25 seedlings showed IP = 0, while others showed IP ≥ 3). +; amplification by Bra012688m, -; no amplification by Bra012688m. A, Resistant allele (*FocBr1/FocBr1* or *FocBr1/focbr1-1*). B, Susceptible allele (*focbr1-1/focbr1-2* or *focbr1-2/focbr1-2*). C, Heterozygous allele (*FocBr1/focbr1-2*). D, No PCR amplification (*focbr1-1/focbr1-1*).

Table 4. Genotype distribution of *FocBr1* in *B. rapa* subspecies.

	Chinese Cabbage (var. *pekinensis*)	Turnip (var. *rapa*)	Pak Choi (var. *chinensis*)	Komatsuna (var. *perviridis*)
FocBr1/FocBr1 or *FocBr1/focbr1-1*	145	30	24	44
FocBr1/focbr1-2	6	5	16	21
focbr1-1/focbr1-1	6	0	0	3
focbr1-2/focbr1-2 or *focbr1-2/focbr1-1*	0	0	0	5
Total	157	35	40	73

3. Discussion

Previously, we identified a resistance gene to *Foc* (*FocBr1*), which is a single dominant gene. In susceptible lines, a 35 kb deletion, which includes *FocBr1*, was found [12], and the susceptible allele was termed *focbr1-1* [14]. A dominant DNA marker (Bra012688m) has been made and the prediction of *Foc* resistance using Bra012688m was consistent with phenotypes of *Foc* resistance confirmed by an inoculation test in inbred lines of Chinese cabbage [12,15]. In this study, we inoculated *Foc* and other formae speciales, *For*, to Chinese cabbage, turnip, pak choi, komatsuna, and chijimina lines, and we found the prediction using Bra012688m was not consistent with phenotype using inoculation test in some lines, especially in komatsuna. To clarify the inheritance pattern of the *R* gene to *For*, we performed linkage analysis. Three F_2 populations derived from crosses between *For* resistant and susceptible lines (*focbr1-1*) showed a 3:1 ratio of resistant to susceptible plants. In contrast, the other three F_2 populations derived from crosses between *For* resistant and susceptible lines (*focbr1-2*) did not show a 3:1 ratio of resistant to susceptible plants; the number of susceptible plants of the F_2 population is smaller than the expected number. However, QTL-seq using these populations identified one causative locus. This could be a difference in the detail of the loss of function; FocBr1 of *focbr1-2* allele might have a weak function against *For* or be susceptible to environmental effects, although the *focbr1-1* allele has completely lost its function. However, as both alleles showed a strong susceptible phenotype and we could not identify any significant difference between these two alleles in *B. rapa* lines, further analysis will be needed to identify this minor difference between *focbr1-1* and *focbr1-2* alleles.

QTL-seq analysis using F_2 populations derived from crosses between *For* resistant and susceptible lines with the *FocBr1* deletion (*focbr1-1*) or without the *FocBr1* deletion (*focbr1-2*) identified the same single causative locus for *For* resistance, which covered *FocBr1*. There was no difference in expression levels of *FocBr1* between resistant and susceptible lines, but there were some amino acid sequence differences between *For* susceptible allele (*focbr1-2*) and resistant alleles (*FocBr1* and *FocBo1*), suggesting that changes of amino acid sequence result in loss of function. Some substitutions were in the LRR region, and these susceptible line-specific amino acid changes may lead to loss of recognition of AVR. This new allele might be useful for identifying the sequence that is important for the interaction between R and AVR proteins. To prove this amino acid sequences change results in loss of function, which might be due to loss of recognition to AVR, further experiments such as making transgenic plants for complementation or loss of function by CRISPR-Cas9 system will be required.

Alternatively, another gene(s) linked to the *FocBr1* locus may work together with *FocBr1* for *For* resistance, because the peak detected by QTL-seq is upstream from the *FocBr1* position in three F_2 populations derived from crosses between *For* resistant and susceptible lines (*focbr1-2*). In *Arabidopsis thaliana*, TIR-NBS-LRR type *Resistance to Ralstonia solanacearum 1* (*RRS1*) and *Resistance to Pseudomonas syringae 4* (*RPS4*) are neighboring genes and both are required for resistance to *Colletotrichum higginsianum*, *Ralstonia solanacearum*, and *Pseudomonas syringae* pv. *tomato* strain DC3000 expressing *avrRps4* [18]. RRS1 encodes a WRKY domain protein as well as a TIR-NBS-LRR protein and works as a "sensor" to detect the effector, and RPS4 works as a "helper" to activate cell death [19,20]. If a similar function is applied to Fusarium yellows resistance, FocBr1 will work as a "sensor" NBS-LRR with other "helper" gene(s), which may be located on a region upstream from *FocBr1*. Resistant and susceptible alleles of "sensor", *FocBr1* and *focbr1-2*, might be able to recognize AVR to greater or lesser degrees, respectively, and other "helper" gene(s) might have different functions between resistant and susceptible lines, resulting in a shift of QTL peak to upstream. Further analyses using plants recombined between QTL peak locus and *FocBr1* gene are required to clarify whether other factor(s) are important for *For* resistance and the *For* infection mechanisms. There is also another possibility that minor QTLs not linked to *FocBr1* locus are important for resistance to *For*.

Using a new DNA marker, focbr1-2m, we screened genotypes of *B. rapa* breeding lines and cultivars. There were six lines that showed weak resistance against *For* and *Foc*, and three of the six lines ("CR-Taiga", "Natsurakuten", and YBCG-TC04) showed heterozygosity of *FocBr1* and *focbr1-2*. However, in the remaining three lines ("CR-Yukiakari", NSI-01, and YBCG-18), we cannot distinguish between the homozygosity of *FocBr1* and the heterozygosity of *FocBr1* and *focbr1-1*, because *focbr1-1* results in deletion of *FocBr1* and focbr1-2m cannot amplify *focbr1-1* allele. In *B. rapa*, plants having a homozygous clubroot resistance gene show more stable clubroot resistance than plants having a heterozygous resistance gene [21–23]. Three lines may be heterozygous for *FocBr1* and *focbr1-1*. In the case of *focbr1-1*, it is desirable to develop a DNA marker to distinguish between *FocBr1*/*focbr1-1* heterozygosity and *FocBr1*/*FocBr1* homozygosity, i.e., using a linked marker close to the 35-kb deletion [14]. In the case of *focbr1-2*, codominant DNA marker, focbr1-2m, can distinguish the heterozygosity of *FocBr1* and *focbr1-2* alleles, which will be useful for breeding stable Fusarium yellows resistant cultivars.

In Chinese cabbage cultivars, there were no cultivars homozygous for *focbr1-2*, and a few lines heterozygous for *FocBr1*/*focbr1-2*. In our previous study using the Bra012688m marker, there was complete agreement between the DNA marker-based prediction and the inoculation test in Chinese cabbage lines [15]. Thus, there is little risk that the presence of the *focbr1-2* allele leads to susceptibility during the breeding of Fusarium yellows resistant cultivars in Chinese cabbage. Like in Chinese cabbage, most turnip cultivars (about 85%) did not have *focbr1-2* alleles, so this allele will not be a problem for breeding. However, in pak choi, 40% of cultivars were heterozygous for *FocBr1* and *focbr1-2* alleles. In komatsuna, about 30% of cultivars were heterozygous for *FocBr1* and *focbr1-2* alleles and about 7% of cultivars were homozygous for *focbr1-2* allele or heterozygous for *focbr1-2* and *focbr1-1* alleles. For the breeding of Fusarium yellows resistant cultivars in pak choi or komatsuna, the presence of the *focbr1-2* allele should be mapped in breeding lines, and the DNA marker, focbr1-2m, developed in this study will be useful for DNA marker-assisted selection.

4. Materials and Methods

4.1. Plant Materials and DNA and RNA Extraction

The breeding lines and commercial F_1 hybrid cultivars of *B. rapa* vegetables were used as plant materials (Table S5). F_2 populations were produced by bud pollination of F_1 hybrid crossing YBCG-11 × YBCG-12, YBCG-11 × YBCG-13, YBCG-11 × YBCG-14, YBCG-08 × YBCG-09, YBCG-TC01 × YBCG-10, and YBCG-11 × YBCG-10. Genomic DNA was isolated from leaves by the CTAB (cetyl trimethyl ammonium bromide) method [24]. Total RNA was isolated from noninoculated leaves of ten-days-old seedlings by the SV Total RNA Isolation System (Promega Co., Madison, WI, USA).

4.2. Inoculation Test

A strain of *F. oxysporum* f. sp. *rapae* (isolated from komatsuna) (provided by NARO, MAFF 240322) or *F. oxysporum* f. sp. *conglutinans* (isolated from cabbage) was used to prepare inocula. Liquid inocula were obtained by inoculating potato sucrose broth medium (200 g/L potato extract and 20 g/L sucrose in distilled water) with the isolate and shaking at 130 rpm on a rotary shaker for 1 week. Roots of ten-days-old seedlings were dipped in fungal spore suspension (fungal titer of ~5×10^6) for 5 h and then transplanted into a cell tray filled with soil. Plants were grown in the greenhouse, and two or three weeks after inoculation, individual plants were scored for interaction phenotype (IP) based on six categories that are 0 (no symptoms in tops and roots), 3 (darkening of roots, slight top stunting, and no chlorosis), 5 (dark stunted roots, tops stunted, and slight chlorosis of cotyledons), 7 (severe stunting of roots and tops and strong chlorosis) and 9 (severe stunting, necrosis, and death). To show the phenotype of breeding lines and cultivars, the average IP among 25 seedlings were categorized into resistant (IP = 0) or susceptible (IP = 3–9). Average IP of most resistant lines/cultivars was around 0 and average IP of most susceptible lines/cultivars was 9. However, some of the 25 seedlings of the line showed IP = 0 while others showed IP $\geq$ 3, and these exceptions are represented by R*. In the linkage analysis, F_2 seedlings were used for inoculation test, and phenotypes of individual seedlings were resistant (IP = 0) or susceptible (IP = 3–9). Chinese cabbage inbred lines RJKB-T23 and RJKB-T24 were used as a resistant and susceptible control [12].

4.3. QTL-Seq

QTL-seq was performed following the method described in [25]. From F_2 populations derived from YBCG-11 × YBCG-12, YBCG-11 × YBCG-13, YBCG-11 × YBCG-14, and YBCG-08 × YBCG-09 hybrids, about 20 plants were selected from resistant (IP = 0) and susceptible (IP = 9) plants based on their perfectly resistance or susceptible phenotype. The equal amount of DNA from each sample was bulked by resistant and susceptible phenotypes, and named R-bulk and S-bulk, respectively. Eight sequence libraries were prepared for DNA sequencing using TruSeq DNA PCR-Free kit (Illumina, Inc., San Diego, CA, USA), and sequenced by Illumina Hiseq 4000 (paired end, 150 bp). For detecting the parental SNPs, DNA from parental line (YBCG-08) was also sequenced.

Sequence reads were quality trimmed by FaQCs. Trimmed reads of R-bulk and S-bulk were aligned to the *B. rapa* reference genome version 3.0 (https://brassicadb.cn, accessed on 1 April 2021), and SNP-index was calculated at all SNPs in R-bulk and S-bulk compared with resistant parental sequences, then the subtracted value of SNP-index of R-bulk from SNP-index of S-bulk was calculated as ΔSNP-index using QTL-seq pipeline.

4.4. Prediction of Fusarium Yellows Resistance by DNA Markers

To predict the Fusarium yellows resistance, the dominant marker Bra012688m [15] and codominant CAPS marker, focbr1-2m, were used. PCR was performed using QuickTaq®HS DyeMix (TOYOBO Co., Ltd., Osaka, Japan). The reaction mixture was incubated in the thermal cycler (TaKaRa PCR Thermal Cycler Dice® Gradient, Takara Bio Inc., Kusatsu, Japan) at 94 °C for 2 min following by 35 cycles of 94 °C for 30 s, 55 °C for 30 s, and 68 °C for 1 min. PCR products were detected by electrophoresis (i-MyRunII, COSMO BIO CO., LTD., Tokyo, Japan) using 1.0% agarose gel (Bra012688m). To distinguish the *FocBr1* and *focbr1-2* alleles, amplified DNA digested by *Hin*d III restriction enzyme were electrophoresed on 1.5% agarose gel. Two or more independent individual plants in each cultivar were tested for genotyping. The primer sets of the DNA markers are listed in Table S6.

4.5. Gene Expression Analysis

cDNA was synthesized from 500 ng total RNA using ReverTra Ace qPCR RT Master Mix with gDNA Remover (TOYOBO Co., Ltd.). The specificity of the primer set of *FocBr1* was first tested by electrophoresis of RT-PCR amplified products using QuickTaq®HS DyeMix on 1.5% agarose gel in which a single product was observed. RT-PCR conditions were 94 °C for 2 min followed by 35 cycles of 94 °C for 30 s, 55 °C for 30 s, and 68 °C for 30 s. The absence of genomic DNA contamination was confirmed by the PCR of the no RT control. To distinguish the *FocBr1* and *focbr1-2* alleles, amplified DNA by RT-PCR digested by *Hin*d III restriction enzyme were electrophoresed on 1.5% agarose gel. The primer sets for RT-PCR are listed in Table S6.

5. Conclusions

In this study, we identified *FocBr1* as a *For* resistance gene, and a new susceptible allele of *FocBr1*, *focbr1-2*, was identified in *B. rapa*. Furthermore, a new DNA marker, which can distinguish between *FocBr1*, *focbr1-1*, and *focbr1-2*, was developed.

Supplementary Materials: The following are available online at https://www.mdpi.com/article/10.3390/plants10061082/s1, Figure S1: QTL-seq results of all chromosomes of F_2 populations derived from YBCG-11 × YBCG-12 hybrid, Figure S2: QTL-seq results of all chromosomes of F_2 populations derived from YBCG-11 × YBCG-13 hybrid, Figure S3: QTL-seq results of all chromosomes of F_2 populations derived from YBCG-11 × YBCG-14 hybrid, Figure S4: Overlapped candidate genes in three F_2 populations, Figure S5: QTL-seq results of all chromosomes of F_2 populations derived from YBCG-08 × YBCG-09 hybrid, Table S1: Assessment of Fusarium yellows resistance by inoculation test, Table S2: Genes located on candidate locus detected by QTL-seq in four populations by 95% significance, Table S3: NBS-LRR encoded genes in significant QTLs, Table S4: Genotype using focbr1-2m marker in resistant and susceptible plants derived from F_2 population, Table S5: List of cultivars used in this study, Table S6: List of primer sequences.

Author Contributions: Conceptualization, N.M., C.S., T.I., K.O., M.S., M.K. (Makoto Kaji), M.K. (Masahiko Katsumata), and R.F.; methodology, validation and formal analysis, N.M., M.A.A., C.S., H.M., T.S., T.I., M.S. and R.F.; writing—original draft preparation, N.M., M.A.A., M.S., E.S.D. and R.F.; writing—review and editing, N.M., E.S.D. and R.F.; visualization, N.M. and M.S.; supervision, M.K. (Masahiko Katsumata) and R.F.; funding acquisition, T.I., M.K. (Makoto Kaji), M.K. (Masahiko Katsumata) and R.F. All authors have read and agreed to the published version of the manuscript.

Funding: This work was funded by grants from the Project of the NARO Bio-oriented Technology Research Advancement Institution (Research program on Development of Innovation Technology) (30029C).

Acknowledgments: We are grateful to Yumiko Arai for her technical assistance throughout this project. We thank Kenji Osabe for his helpful comments and manuscript editing.

Conflicts of Interest: The authors declare no conflict of interest.

References

1. Cheng, F.; Sun, R.; Hou, X.; Zheng, H.; Zhang, F.; Zhang, Y.; Liu, B.; Liang, J.; Zhuang, M.; Liu, Y.; et al. Subgenome parallel selection is associated with morphotype diversification and convergent crop domestication in *Brassica rapa* and *Brassica oleracea*. *Nat. Genet.* **2016**, *48*, 1218–1224. [CrossRef] [PubMed]
2. Lv, H.; Miyaji, N.; Osabe, K.; Akter, A.; Mehraj, H.; Shea, D.J.; Fujimoto, R. The importance of genetic and epigenetic research in the *Brassica* vegetables in the face of climate change. In *Genomic Designing of Climate-Smart Vegetable Crops*; Kole, C., Ed.; Springer: Berlin/Heidelberg, Germany, 2020; pp. 161–255.
3. Fujimoto, R.; Uezono, K.; Ishikura, S.; Osabe, K.; Peacock, W.J.; Dennis, E.S. Recent research on the mechanism of heterosis is important for crop and vegetable breeding systems. *Breed. Sci.* **2018**, *68*, 145–158. [CrossRef] [PubMed]
4. Akter, A.; Itabashi, E.; Kakizaki, T.; Okazaki, K.; Dennis, E.S.; Fujimoto, R. Genome triplication leads to transcriptional divergence of *FLOWERING LOCUS C* genes during vernalization in the genus *Brassica*. *Front. Plant Sci.* **2021**, *11*, 619417. [CrossRef] [PubMed]
5. Mehraj, H.; Akter, A.; Miyaji, N.; Miyazaki, J.; Shea, D.J.; Fujimoto, R.; Doullah, M.A.U. Genetics of clubroot and Fusarium wilt disease resistance in Brassica vegetables: The application of marker assisted breeding for disease resistance. *Plants* **2020**, *9*, 726. [CrossRef] [PubMed]
6. Lv, H.; Fang, Z.; Yang, L.; Zhang, Y.; Wang, Y. An update on the arsenal: Mining resistance genes for disease management of *Brassica* crops in the genomic era. *Hortic. Res.* **2020**, *7*, 34. [CrossRef] [PubMed]
7. Jones, J.D.G.; Dangl, J.L. The plant immune system. *Nature* **2006**, *444*, 323–329. [CrossRef] [PubMed]
8. Dodds, P.N.; Rathjen, J.P. Plant immunity: Towards an integrated view of plant-pathogen interactions. *Nat. Rev. Genet.* **2010**, *11*, 539–548. [CrossRef]
9. Enya, J.; Togawa, M.; Takeuchi, T.; Yoshida, S.; Tsushima, S.; Arie, T.; Sakai, T. Biological and phylogenetic characterization of *Fusarium oxysporum* complex, which causes yellows on *Brassica* spp., and proposal of *F. oxysporum* f. sp. *rapae*, a novel forma specialis pathogenic on *B. rapa* in Japan. *Phytopathology* **2008**, *98*, 475–483. [CrossRef]
10. Pu, Z.; Shimizu, M.; Zhang, Y.; Nagaoka, T.; Hayashi, T.; Hori, H.; Matsumoto, S.; Fujimoto, R.; Okazaki, K. Genetic mapping of a fusarium wilt resistance gene in *Brassica oleracea*. *Mol. Breed.* **2012**, *30*, 809–818. [CrossRef]
11. Lv, H.; Fang, Z.; Yang, L.; Zhang, Y.; Wang, Q.; Liu, Y.; Zhuang, M.; Yang, Y.; Xie, B.; Liu, B.; et al. Mapping and analysis of a novel candidate Fusarium wilt resistance gene *FOC1* in *Brassica oleracea*. *BMC Genom.* **2014**, *15*, 1094. [CrossRef]

12. Shimizu, M.; Fujimoto, R.; Ying, H.; Pu, Z.; Ebe, Y.; Kawanabe, T.; Saeki, N.; Taylor, J.M.; Kaji, M.; Dennis, E.S.; et al. Identification of candidate genes for fusarium yellows resistance in Chinese cabbage by differential expression analysis. *Plant Mol. Biol.* **2014**, *85*, 247–257. [CrossRef]
13. Shimizu, M.; Pu, Z.; Kawanabe, T.; Kitashiba, H.; Matsumoto, S.; Ebe, Y.; Sano, M.; Funaki, T.; Fukai, E.; Fujimoto, R.; et al. Map-based cloning of a candidate gene conferring Fusarium yellows resistance in *Brassica oleracea*. *Theor. Appl. Genet.* **2015**, *128*, 119–130. [CrossRef]
14. Akter, M.A.; Mehraj, H.; Itabashi, T.; Shindo, T.; Osaka, M.; Akter, A.; Miyaji, N.; Chiba, N.; Miyazaki, J.; Fujimoto, R. Breeding for disease resistance in *Brassica* vegetables using DNA marker selection. In *Brassica Breeding and Biotechnology*; IntechOpen: London, UK, 2021. [CrossRef]
15. Kawamura, K.; Kawanabe, T.; Shimizu, M.; Okazaki, K.; Kaji, M.; Dennis, E.S.; Osabe, K.; Fujimoto, R. Genetic characterization of inbred lines of Chinese cabbage by DNA markers: Towards the application of DNA markers to breeding of F_1 hybrid cultivars. *Data Brief* **2016**, *6*, 229–237. [CrossRef] [PubMed]
16. Kawamura, K.; Shimizu, M.; Kawanabe, T.; Pu, Z.; Kodama, T.; Kaji, M.; Osabe, K.; Fujimoto, R.; Okazaki, K. Assessment of DNA markers for seed contamination testing and selection of disease resistance in cabbage. *Euphytica* **2017**, *213*, 28. [CrossRef]
17. Sato, M.; Shimizu, M.; Shea, D.J.; Hoque, M.; Kawanabe, T.; Miyaji, N.; Fujimoto, R.; Fukai, E.; Okazaki, K. Allele specific DNA marker for fusarium resistance gene *FocBo1* in *Brassica oleracea*. *Breed. Sci.* **2019**, *69*, 308–315. [CrossRef] [PubMed]
18. Narusaka, M.; Shirasu, K.; Noutoshi, Y.; Kubo, Y.; Shiraishi, T.; Iwabuchi, M.; Narusaka, Y. *RRS1* and *RPS4* provide a dual *Resistance*-gene system against fungal and bacterial pathogens. *Plant J.* **2009**, *60*, 218–226. [CrossRef]
19. Huh, S.U.; Cevik, V.; Ding, P.; Duxbury, Z.; Ma, Y.; Tomlinson, L.; Sarris, P.F.; Jones, J.D.G. Protein-protein interactions in the RPS4/RRS1 immune receptor complex. *PLoS Pathog.* **2017**, *13*, e1006376. [CrossRef]
20. Ma, Y.; Guo, H.; Hu, L.; Martinez, P.P.; Moschou, P.N.; Cevik, V.; Ding, P.; Duxbury, Z.; Sarris, P.F.; Jones, J.D.G. Distinct modes of derepression of an *Arabidopsis* immune receptor complex by two different bacterial effectors. *Proc. Natl. Acad. Sci. USA* **2018**, *115*, 10218–10227. [CrossRef]
21. Suwabe, K.; Tsukazaki, H.; Iketani, H.; Hatakeyama, K.; Fujimura, M.; Nunome, T.; Fukuoka, H.; Matsumoto, S.; Hirai, M. Identification of two loci for resistance to clubroot (*Plasmodiophora brassicae* Woronin) in *Brassica rapa* L. *Theor. Appl. Genet.* **2003**, *107*, 997–1002. [CrossRef]
22. Nomura, K.; Minegishi, Y.; Kimizuka-Takagi, C.; Fujioka, T.; Moriguchi, K.; Shishido, R.; Ikehashi, H. Evaluation of F_2 and F_3 plants introgressed with QTLs for clubroot resistance in cabbage developed by using SCAR markers. *Plant Breed.* **2005**, *124*, 371–375. [CrossRef]
23. Nagaoka, T.; Doullah, M.A.U.; Matsumoto, S.; Kawasaki, S.; Ishikawa, T.; Hori, H.; Okazaki, K. Identification of QTLs that control clubroot resistance in *Brassica oleracea* and comparative analysis of clubroot resistance genes between *B. rapa* and *B. oleracea*. *Theor. Appl. Genet.* **2010**, *120*, 1335–1346. [CrossRef] [PubMed]
24. Murray, M.G.; Thompson, W.F. Rapid isolation of high molecular weight plant DNA. *Nucleic Acids Res.* **1980**, *8*, 4321–4325. [CrossRef] [PubMed]
25. Takagi, H.; Abe, A.; Yoshida, K.; Kosugi, S.; Natsume, S.; Mitsuoka, C.; Uemura, A.; Utsushi, H.; Tamiru, M.; Takuno, S.; et al. QTL-seq: Rapid mapping of quantitative trait loci in rice by whole genome resequencing of DNA from two bulked populations. *Plant J.* **2013**, *74*, 174–183. [CrossRef] [PubMed]

Article

Mapping of Genetic Locus for Leaf Trichome Formation in Chinese Cabbage Based on Bulked Segregant Analysis

Rujia Zhang [†], Yiming Ren [†], Huiyuan Wu, Yu Yang, Mengguo Yuan, Haonan Liang and Changwei Zhang *

State Key Laboratory of Crop Genetics and Germplasm Enhancement, College of Horticulture, Nanjing Agricultural University, Nanjing 210095, China; 2018204020@njau.edu.cn (R.Z.); 2018104075@njau.edu.cn (Y.R.); 2019104075@njau.edu.cn (H.W.); 2019804226@njau.edu.cn (Y.Y.); 2020104077@njau.edu.cn (M.Y.); 2020804243@njau.edu.cn (H.L.)
* Correspondence: changweizh@njau.edu.cn; Tel.: +86-025-84395332
† These authors equally contributed to the work.

Citation: Zhang, R.; Ren, Y.; Wu, H.; Yang, Y.; Yuan, M.; Liang, H.; Zhang, C. Mapping of Genetic Locus for Leaf Trichome Formation in Chinese Cabbage Based on Bulked Segregant Analysis. *Plants* **2021**, *10*, 771. https://doi.org/10.3390/plants10040771

Academic Editors: Ioannis Ganopoulos and Yoshinobu Takada

Received: 16 March 2021
Accepted: 13 April 2021
Published: 14 April 2021

Publisher's Note: MDPI stays neutral with regard to jurisdictional claims in published maps and institutional affiliations.

Copyright: © 2021 by the authors. Licensee MDPI, Basel, Switzerland. This article is an open access article distributed under the terms and conditions of the Creative Commons Attribution (CC BY) license (https://creativecommons.org/licenses/by/4.0/).

Abstract: Chinese cabbage is a leafy vegetable, and its leaves are the main edible organs. The formation of trichomes on the leaves can significantly affect its taste, so studying this phenomenon is of great significance for improving the quality of Chinese cabbage. In this study, two varieties of Chinese cabbage, W30 with trichome leaves and 082 with glabrous leaves, were crossed to generate F_1 and F_1 plants, which were self-fertilized to develop segregating populations with trichome or glabrous morphotypes. The two bulks of the different segregating populations were used to conduct bulked segregant analysis (BSA). A total of 293.4 M clean reads were generated from the samples, and plants from the trichome leaves (AL) bulk and glabrous leaves (GL) bulk were identified. Between the two DNA pools generated from the trichome and glabrous plants, 55,048 SNPs and 272 indels were generated. In this study, three regions (on chromosomes 6, 10 and scaffold000100) were identified, and the annotation revealed three candidate genes that may participate in the formation of leaf trichomes. These findings suggest that the three genes—*Bra025087* encoding a cyclin family protein, *Bra035000* encoding an ATP-binding protein/kinase/protein kinase/protein serine/threonine kinase and *Bra033370* encoding a WD-40 repeat family protein–influence the formation of trichomes by participating in trichome morphogenesis (GO: 0010090). These results demonstrate that BSA can be used to map genes associated with traits and provide new insights into the molecular mechanism of leafy trichome formation in Chinese cabbage.

Keywords: bulked segregant analysis; Chinese cabbage; leaf trichome

1. Introduction

Chinese cabbage (*Brassica rapa* ssp. *pekinensis*), one of the subspecies of *Brassica rapa* [1], is one of the most popular vegetable crops in Asia, and its leaves are the main edible organs. Trichome formation in Chinese cabbage during its growth and development may affect how it tastes to consumers. The formation of leaves' trichomes is receiving increasing attention from researchers, but the literature on its molecular mechanism remains insufficient.

Trichomes play important roles in water regulation, temperature control and the protection of plants against biotic and abiotic stresses, thereby increasing their tolerance to changes in the environment [2]. However, if the edible part is covered with trichomes, it may influence the appearance and mouthfeel [3,4]. This has caused researchers to attach greater importance to the molecular mechanism of the formation of trichomes. Trichomes arise at an early stage of organ morphogenesis out of the epidermal progenitor cells that also give rise to other cell types such as stomata and pavement cells. Trichomes are single-celled and hairy structures that develop on the epidermis of the aerial parts, including leaves, stems, fruits and sepals. Previous studies have reported the molecular mechanism of trichome formation in other plants, such as Arabidopsis [5], rice [6], cotton [7], cucumber [8], tomato [9] and maize [10]. However, few studies have been reported on trichome formation in Chinese cabbage [11].

Bulked segregant analysis (BSA) is a simple and rapid approach that uses segregating populations to identify molecular markers that are tightly linked to the causal gene underlying a given phenotype [12,13]. Two bulked pools with extreme traits and two parental lines were constructed for high-throughput next-generation sequencing (NGS) to identify polymorphic markers and conduct correlation analysis. The aim of the latter is to annotate the functions of the genes in the mapped regions by aligning the reference genome sequence of the species. With the development of DNA sequencing technology, NGS-based BSA has been used to map important genes in many plants, such as the yellow rind formation gene in watermelon [14], the branching habit trait in cultivated peanut [15], a candidate nicosulfuron sensitivity gene in maize [16], grain-shape-related loci in rice [17] and genes associated with the heading type of Chinese cabbage [18]. A large number of studies have proven that bulked segregant analysis is a reliable method for identifying loci associated with traits in plants.

In this study, we applied BSA-Seq to identify the pathways and genes related to the formation of trichomes. New SNPs and indels were developed to perform fine-linkage mapping of the previously located region. Taken together, the results provide new insights into the fine-mapping and identification of candidate genes in horticultural crops.

2. Results

2.1. Morphology and Genetic Analysis of Hairy and Hairless Chinese Cabbage Plant

There were significant differences in the surfaces of leaves between the W30 and 082 varieties of Chinese cabbage, which was further confirmed by integrated microscopy. F_1 plants from a cross between W30 and 082 displayed trichome leaves. When the F_1 plants were self-fertilized, the F_2 plants showed different phenotypes: some had trichome leaves, and others had glabrous leaves (Figure 1). After sowing the F_2 generation, a total of 294 plants were grown to observe whether the leaves formed trichomes. Among these plants, 212 developed glandular trichomes, while 82 plants were observed with glabrous leaves. This corresponds to a three to one segregation ratio (Table 1). These results demonstrate that trichome formation in Chinese cabbage is controlled by a dominant gene.

2.2. Construction and Sequencing of the Trichome Leaves (AL) Bulk and Glabrous Leaves (GL) Bulk Samples and Parental Lines

In order to further explore the candidate genes that regulate the formation of trichomes, we used the BSA-Seq strategy to identify candidate regions and find genes related to the formation of trichomes. Fifty plants with glandular trichomes and fifty plants with glabrous leaves were randomly selected from the F_2 population, which contained a mixture of the AL-bulk and GL-bulk, and sequenced with their parental lines. After screening to remove low-quality reads, the two parents were resequenced, resulting in 37,948,696 and 36,842,550 reads and 11.38 and 11.05 Gb from W30 and 082, respectively. A total of 65.57 Gb clean data were obtained from the two bulks (33.06 Gb for the AL-bulk and 32.51 Gb for the GL-bulk). After mapping these reads to the reference genome, the coverage of AL and GL genomes was found to be 78× and 76×, respectively (Table 2).

Table 1. Genetic analysis of leaf trichome phenotype.

Generation	Trichome Leaves	Glabrous Leaves	Segregation Ratio
P1(W30)	40	0	
P2(082)	0	40	
F1	60	0	
F2	212	82	3:1

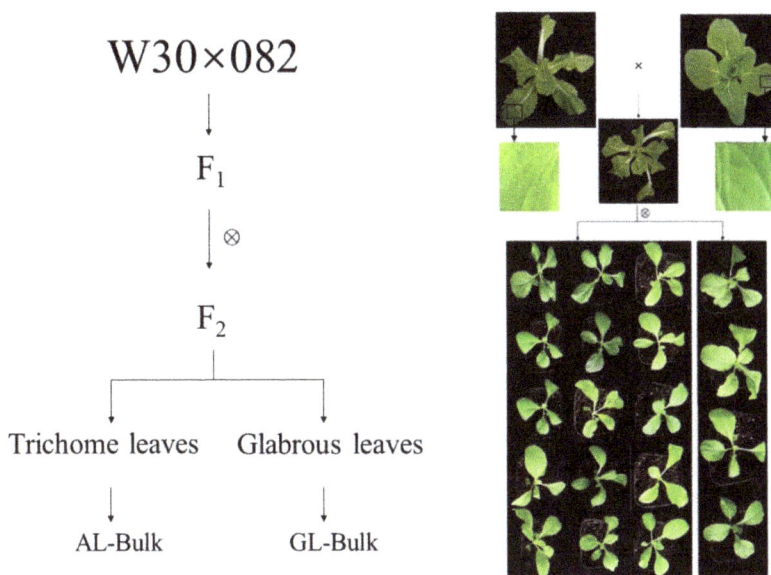

Figure 1. The phenotypes of plants of the W30 and 082 parental lines and their F_1 and F_2 populations. W30 had trichome leaves, and 082 had glabrous leaves. All F1 plants had trichomes and the leaves of the F2 population were classified as either trichome or glabrous.

Table 2. Quality control of sequencing data for parental inbred lines W30 and 082 and bulked AL-pool and GL-pool samples.

Bulk	Clean Reads	Data Generated	Q30 (30%)	Genome Coverage (10×)	Average Depth (×)	SNP Number	Alignment Efficiency (%)
W30	37,948,696	11,384,608,800	92.89	93.43	27.5384	1,693,338	97
082	36,842,550	11,052,765,000	93.55	92.44	26.8289	1,663,130	97.26
AL	110,197,965	33,059,389,500	92.69	97.74	78.9208	1,740,465	97.28
GL	108,389,140	32,516,742,000	92.66	97.49	76.169	1,721,183	96.68

2.3. Selection of Candidate Regions

Between AL and GL bulks, 3581 SNP sites were screened according to the principle of genotypic inconsistency (inconsistency between samples or heterozygous SNP sites) in the progeny mixing pool (AL and GL), and the depth was no less than 5X. To identify the genomic region associated with the trichome formation, we used the SNP-index to measure the allele segregation of SNPs between the two bulks. The method used to calculate the Δ(SNP-index) was in accordance with Yuanting [19]. In the method for determining the Δ(SNP-index) threshold of the nonreference genome, a Δ(SNP-index) greater than 0.99 was selected as the threshold for defining significant associations of a marker with traits; that is, a marker larger than the threshold was deemed to be significantly associated with traits. Six significant regions associated with trichome formation were detected by Δ(SNP-index) analysis (Figure 2A). They were located on Scaffold000100 from 160,071 to 260,071, Scaffold001011 from −49,220 to 50,780, Scaffold004266 from −49,807 to 50,193, Scaffold000169 from 132,175 to 132,181 and chromosome 6 from 22,044,767 to 22,246,745 and 23,704,762 to 24,097,454 bp. The identified regions contain a total of 894,682 bp and 110 genes were contained (Table 3).

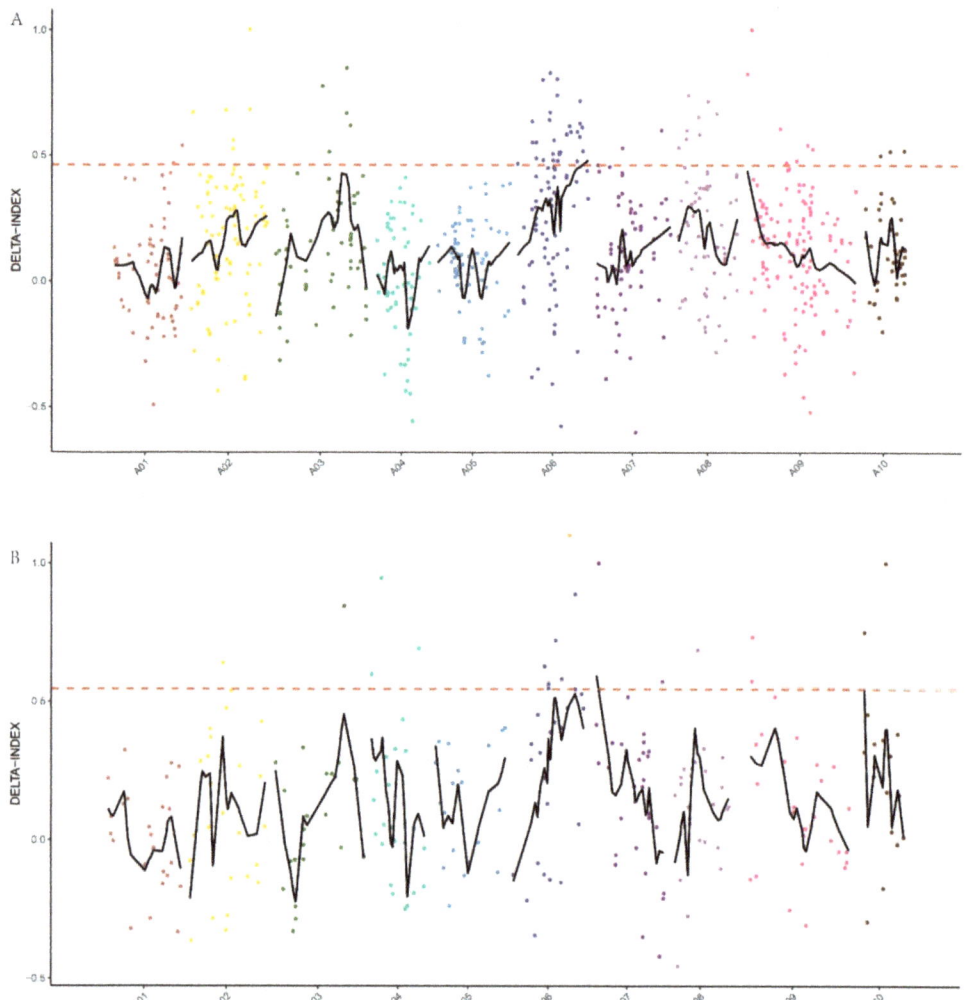

Figure 2. Candidate genomic regions for trichome formation identified using (**A**) the SNP-index algorithm and (**B**) the indel-index algorithm. The red dashed horizontal lines represent the threshold for defining a significant association. The X-axis shows the chromosome position and the different colors represent the different chromosomes. The Y-axis represents the Δ-index values.

Table 3. SNP-select region information.

Chrom	Start	End	Length	Number of Genes
Scaffold000100	160,071	260,071	100,001	6
Scaffold001011	−49,220	50,780	100,001	0
Scaffold004266	−49,807	50,193	100,001	0
A06	22,044,767	22,246,745	201,979	30
A06	23,704,762	24,097,454	392,693	74
Scaffold000169	132,175	132,181	7	0

According to the principle of genotype inconsistency (inconsistency between samples or heterozygous indel sites) in the progeny mixing pool (AL and GL), the depth was no less than 5X, and 1131 indel sites were screened. Using the indel data, three candidate regions were identified (Figure 2B). These regions are located on chromosome 7 from 1,892,096 to 1,992,096, chromosome 10 from 2,673,013 to 2,773,013 and scaffold000100 from 764,907 to 864,907. These candidate regions have a total length of 30,003 bp and contain a total of 45 genes (Table 4).

Table 4. Indel-select region information.

Chrom	Start	End	Length	Number of Genes
A07	1,892,096	1,992,096	100,001	13
A10	2,673,013	2,773,013	100,001	23
Scaffold000100	764,907	864,907	100,001	9

2.4. Gene Ontology (GO) Classification Analysis of Candidate Genes

A GO classification analysis was carried out to understand the functions of all the candidate genes identified in the association analysis of SNPs and small indels. Comparing the AL-bulk with the GL-bulk revealed a total of 108 candidate genes in the analysis of SNPs, and 44 candidate genes were identified by analyzing the effects of small indels through GO annotation (Figure 3). All candidate genes are divided into three categories: biological processes, cellular components, and molecular functions. In general, the genes in candidate regions are more abundant in biological process classes than all other genetic background classes, including the other two categories identified in the analysis. Biological processes include cellular processes, metabolic processes, single-cell processes, response to stimulus, biological regulation, etc., indicating that the formation of leaf trichomes may affect the development and biological processes (Tables S1 and S2).

Three candidate genes (*Bra025087*, *Bra035000* and *Bra033370*) that may be associated with leaf trichome formation in Chinese cabbage were identified through functional annotation (Tables S1 and S2). Table 5 shows the functional annotation of the three candidate genes. To analyze the functions of the candidate genes in the leaf trichome formation, qRT-PCR analysis was performed on W30, 082, five F_1 plants, two F_2 plants with glabrous leaves and six F_2 plants with trichome leaves. According to the results of qRT-PCR analysis, the expression levels of two genes (*Bra025087* and *Bra033370*) were significantly higher in W30, F_1 and F_2 with leaf trichomes than in 082 and F_2 with glabrous leaves (Figure 4A,B). On the contrary, the expression level of *Bra035000* was lower in W30, F_1 and F_2 with leaf trichomes than in 082 and F_2 with glabrous leaves (Figure 4C). This indicates that the two genes (*Bra025087* and *Bra033370*) may facilitate the formation of leaf trichomes, whereas *Bra035000* may suppress it.

2.5. Candidate Genes for Hairiness

Table 5. Three genes related to trichome formation with function annotation by bulked segregant analysis (BSA).

Gene	Chr.	Function Annotation
Bra025087	A10	Cyclin family protein
Bra035000	Scaffold000100	ATP-binding/kinase/protein kinase/protein serine/threonine kinase
Bra033370	A06	WD-40 repeat family protein/beige-related

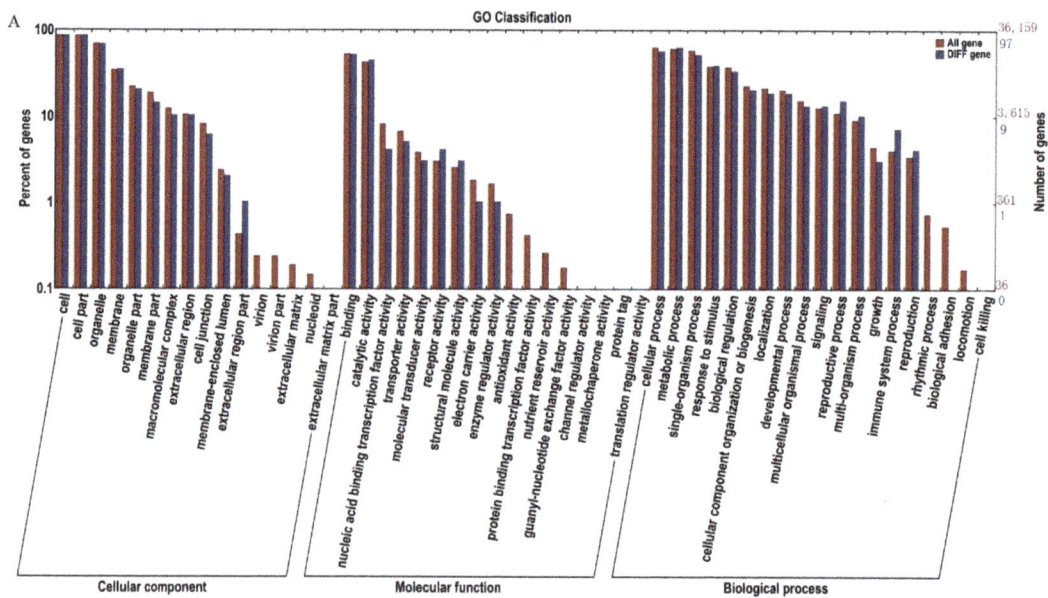

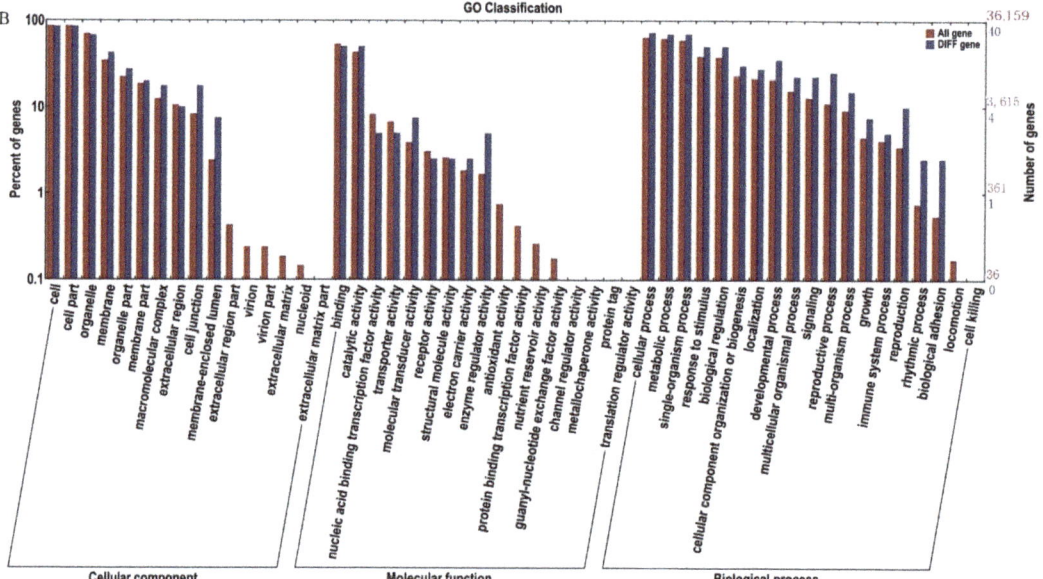

Figure 3. Functional enrichment analysis based on candidate genes GO classification for the (**A**) SNP-selected region and (**B**) the indel-selected region.

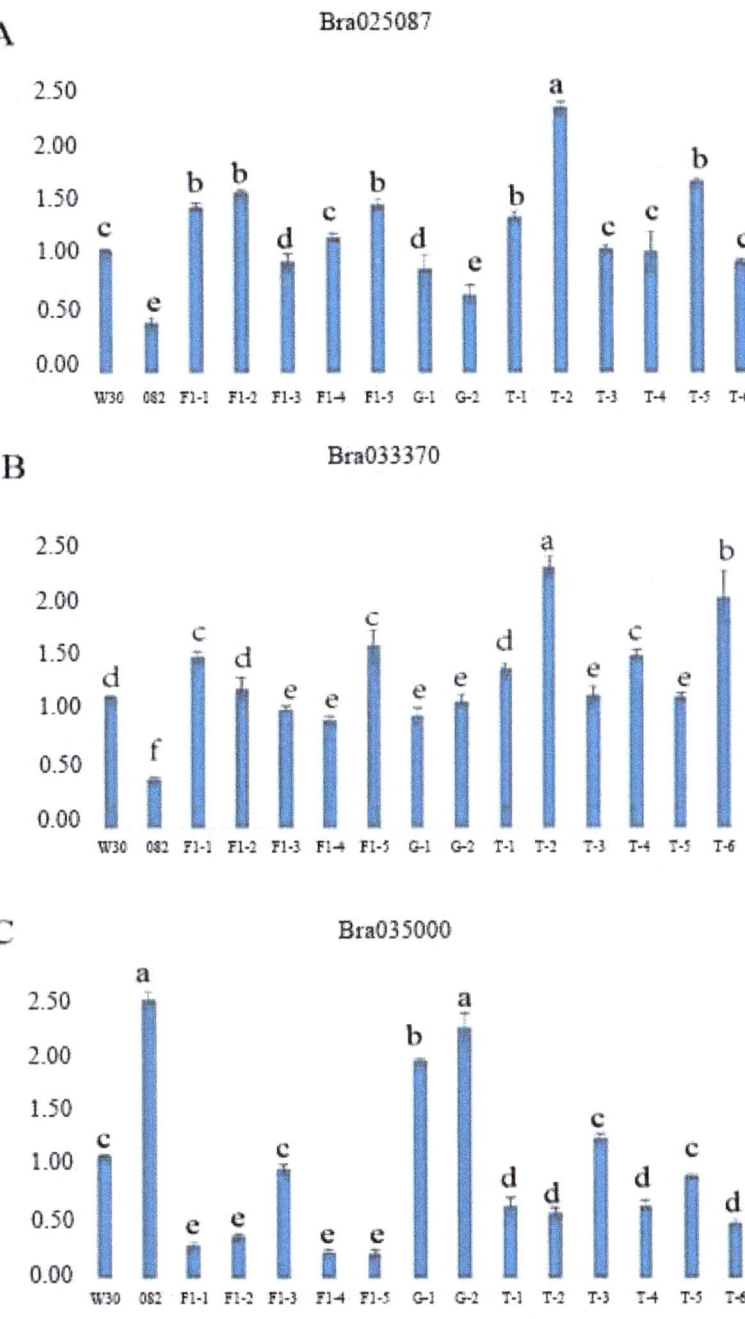

Figure 4. qRT-PCR expression analysis of the candidate genes in the parental lines and F_1 and F_2 generations. (**A**) *Bra025087*; (**B**) *Bra033370*; (**C**) *Bra035000*. W30: female parent; 082: male parent; F_1: 082 × W30; G: F_2 with glabrous leaves, T: F_2 with trichome leaves. The data are the mean + SD; the letters above the bars indicate the significant differences determined by Tukey's honestly significant difference method ($p < 0.05$).

3. Discussion

Although there are many methods for analyzing trait-gene association analysis [20–22], BSA-Seq is still one of the most popular methods and used extensively for various taxa [18,23–26]. The greatest advantage of BSA-Seq is its simplicity in terms of both sample collection and data analysis. Bulked segregant RNA-Seq (BSR-Seq), which combines RNA-Seq with BSA-Seq, is an efficient method for reducing genome complexity [27]. However, if RNA-Seq is not performed using the right tissue at the appropriate developmental stage, the results of BSR-Seq can be misleading. In contrast, samples for BSA-Seq can be collected at any developmental stage and from any tissue. In addition, there is a high degree of structural variation within *Brassica rapa*, which will have a greater negative influence on the data analysis of BSR-Seq.

The key goal of BSA-Seq is to define the smallest candidate regions that are associated with the phenotype. Here, the candidate regions discovered by calculating the Δ(SNP-index) have a total length of 30,003 bp and contain a total of 45 genes. The three candidate genes (*Bra025087*, *Bra035000* and *Bra033370*) selected from the candidate regions are associated with leaf trichome formation in Chinese cabbage, as determined by gene functional annotation [28]. The candidate gene *Bra025087* is an ortholog of Arabidopsis CYCT 1;5 (*At5g45190*). The protein sequence of *Bra025087* exhibits high homology with that of AtCYCT1; 5 (Figure S1). CYCT 1;5 is a cyclin gene and classified as a T-type cyclin [29,30]. Arabidopsis contains five genes encoding cyclin T-like proteins (CYCT1;1 to CYCT1;5). In Arabidopsis, the expression of *CYCT1;5* in the anther and the inflorescence is slightly higher than that in other tissues [31]. It has been confirmed that CYCT1;5 induces complete resistance to CaMV, as well as altered leaf and flower growth, and delayed flowering. The article also reported that the adaxial trichomes of *CYCT1;5* RNAi plant leaves have two branches instead of the three branches found in typical trichomes of wild-type leaves [32]. The results of the RT-qPCR suggest that the expression of *Bra025087* is lower in all plants with glabrous leaves (Figure 4A). It is rational to speculate that the phenotype of BrCYCT1;5 mutant is consistent with *AtCYCT1;5* RNAi. BrCYCT1;5 (*Bra025087*) might positively regulate the growth and development of leaf trichome branches. The candidate gene *Bra035000* is an ortholog of Arabidopsis NIMA-related kinase 6 (NEK6, *At3G44200*), which regulates microtubule organization during anisotropic cell expansion. The similarity percentage between the protein sequences of Bra035000 and AtNEK6 is 80.41% (Figure S2). Previous studies have shown that Arabidopsis NEK6 regulates directional cell expansion through the depolymerization of cortical microtubules during interphase [33–37]. Takatani et al. found that NEK6-1 mutants exhibited wavy growth patterns in the fast-growing region of the hypocotyl, and their hypocotyls did not grow straight [38]. Our research also shows that plants with leaf trichomes have higher expression levels in all three generations (Figure 4B). *NEK6* was shown to be involved in the negative regulation of cell differentiation, further suppressing the development of leaf trichomes. *Bra033370*, another candidate gene, is an ortholog of *Arabidopsis SPI (At1g03060)*, and its protein sequence shares 90.98% similarity with that of *At1g03060* (Figure S3). A decrease in the complexity of epidermal pavement cells and curled trichomes was observed in *SPI* mutants of Arabidopsis [39,40]. Thus, AtSPI positively regulates the normal growth of trichomes in Arabidopsis. *Arabidopsis thaliana*, as a fully sequenced model organism, is an excellent model system for studying cell differentiation in structures such as trichomes [41]. However, there is almost no related research on the three candidate genes in *Brassica rapa*. Therefore, the Arabidopsis model for studying trichomes can serve as a suitable reference for this process in *Brassica rapa*. However, we do not yet clearly know how the three candidate genes work in *Brassica rapa*, and further studies are needed.

4. Materials and Methods

4.1. Plant Materials and Phenotyping for Trichomes

W30, with trichome leaves, and nonheading Chinese cabbage 082 (082), with glabrous leaves, were obtained from the Chinese Academy of Agricultural Sciences of Nanjing Agricultural University. W30 and 082 were crossed, and confirmed F1 plants were self-fertilized to develop segregating populations. Then, 294 seeds of F2 were grown on the experimental farm of Nanjing Agricultural University in September of 2017. Two months later, the surviving plants displayed two contrasting trichome phenotypes with trichome (212) and glabrous (82) leaves. From these, we selected 50 plants of each morphotype and used them to establish R01 and R02 bulks for BSA-Seq analysis (Figure 1). Individual W30 and 082 plants were used to establish parental pools, and BSA-Seq was performed on three plants from each parent.

4.2. BSA-Seq and Sequence Alignment

A total of 100 plants (50 with trichome leaves and 50 with glabrous leaves) were selected from the F2 population for bulking. Two DNA pools were constructed by mixing equal amounts of DNA from 50 trichome leaves (AL-pool) and 50 glabrous leaves (GL-pool). DNA samples from the two bulks and two parental lines were prepared according to the standard Illumina protocol to construct sequencing libraries, which were sequenced on an Illumina HiSeq™2500 platform (Illumina, San Diego, CA, USA). Illumina Casava 1.8 was used for cleaning and filtering reads [42]. After low-quality and short reads were filtered out, the filtered short reads of each pool were mapped onto the Chinese cabbage reference genome sequence V1.5 (http://brassicadb.org/brad/ accessed on 30 March 2018) by BWA [43]. SNP calling was followed by GATK Best-Practices [44]. Both GATK and SAM tools were used to detect SNPs to ensure the accuracy of SNPs. SAM tools were used to remove duplicates and mask the effects of PCR duplication [45]. The obtained SNPs and small indels were annotated and predicted using SnpEff software [46], and only the high-quality SNPs with a minimum sequence read depth of five were used for BSA-Seq analysis.

4.3. Mapping of Candidate Genomic Regions by Association Analysis

Heterozygous and inconsistent SNPs (coverage depth >5×) between two contrasting F_2 pools (AL and GL) were selected to calculate the Δ(SNP-index) values, and sliding-window analysis was performed for the association analysis. LOESS regression was performed for Δ(SNP-index) on the same chromosome to obtain the associated threshold. The candidate genomic regions were identified with an average p-value of $p \leq 0.01$.

The Δ(SNP-index) of each locus was calculated with the formula, Δ(SNP-index) = Mindex−Windex, where m and n are ratios of accessions that exhibit the same bases as the hairy parents in the F2 hairy group and in the F2 glabrous group, respectively. A sliding-window analysis was applied to generate Δ(SNP-index) plots with a window size of 5 SNPs and increment of 2 SNPs. Significant high SNP-index values (the 0.1% in the right tail) were identified as the empirical thresholds, where the value is 0.525.

4.4. Gene Annotation in Candidate Regions

In order to annotate genes in candidate regions and analyze their functions, BLAST was used to compare the genes in the associated region with functional databases including NR [47], SwissProt [47], GO [48], COG [49] and KEGG [50].

4.5. RNA Isolation and qRT-PCR Analysis

The total RNA of W30, 082 and F2 with different phenotypes (trichome and glabrous leaves) was extracted using an RNAprep Pure Plant Kit (TianGen Biotech, Beijing, China). The quality and quantity of RNA were assessed using a Nanodrop 2000. Two micrograms of total RNA were reverse transcribed using Hifair®® III 1st Strand cDNA Synthesis SuperMix for qPCR (gDNA digester plus) (Yeasen, Shanghai, China) following the recommended protocol. Products used as templates for qRT-PCR were diluted 5 times with ddH2O. Hieff®® qPCR SYBR Green Master Mix (Yeasen, Shanghai, China) was employed to identify target gene expression, using a Fluorescent Quantity PCR Detectiong System (Bio-Rad I Cycler iQ5 Hercules, Foster City, CA, USA), in accordance with the manufacturer's protocol. Relative gene transcript levels were determined using the method of the comparative threshold cycle (Ct) method with StepOne v.2.02 software installed in the real-time PCR system, and the 2-ΔΔCT method was used to measure the relative expression level measurement normalized to the internal control gene *BrActin* (*Bra028615*, [51]). The primer pairs were designed using GenScript (https://www.genscript.com/tools/real-time-pcrtaqman-primer-design-tool, accessed on 30 October 2020) and are listed in Table S3.

4.6. Statistical Analyses

We employed Tukey's honestly significant difference (HSD) test to determine statistical significance. The difference was considered significant at $p < 0.05$.

5. Conclusions

In this study, three genes associated with leaf trichome formation were identified and verified in Chinese cabbage by sequencing-based bulked segregant analysis. As high-quality genomes have become more widely available, the BSA method has become an increasingly important tool for the rapid mapping of monogenic traits in diverse *Brassica* species. The results of this study suggest that BSA sequencing is valuable for the assisted-selective breeding of Chinese cabbage.

Supplementary Materials: The following are available online at https://www.mdpi.com/article/10.3390/plants10040771/s1, Figure S1. Sequence alignment of CYCT1;5 proteins in *Brassica rapa* and *Arabidopsis thaliana*. The white words with blue background represent *Brassica rapa* and *Arabidopsis thaliana* have the same amino acid. The black words with light blue background represent *Brassica rapa* and *Arabidopsis thaliana* have the different amino acids. Figure S2. Sequence alignment of NEK6 proteins in *Brassica rapa* and *Arabidopsis thaliana*. The white words with blue background represent *Brassica rapa* and *Arabidopsis thaliana* have the same amino acid. The black words with light blue background represent *Brassica rapa* and *Arabidopsis thaliana* have the different amino acids. Figure S3. Sequence alignment of SPI proteins in *Brassica rapa* and *Arabidopsis thaliana*. The white words with blue background represent *Brassica rapa* and *Arabidopsis thaliana* have the same amino acid. The black words with light blue background represent *Brassica rapa* and *Arabidopsis thaliana* have the different amino acids. Table S1. SNP-select region gene annotation, Table S2. Indel-select region gene annotation, Table S3 Primers used for qRT-PCR analysis.

Author Contributions: Conceiving and designing the research, C.Z.; performing the experiments and data analyses, R.Z. and Y.R.; writing—original draft preparation, R.Z.; writing—review and editing, C.Z. and Y.R.; contribution to the experimental design and coordination of the study, H.W. and Y.Y.; contribution to the interpretation of the results and supervision of the study, M.Y. and H.L. All authors have read and agreed to the published version of the manuscript.

Funding: This work was supported by grants from the Nature Science Foundation of Jiangsu Province (BK20191308).

Institutional Review Board Statement: Not applicable.

Informed Consent Statement: Not applicable.

Data Availability Statement: The available data are presented in the manuscript.

Conflicts of Interest: The authors declare no conflict of interest.

References

1. Song, X.; Ge, T.; Li, Y.; Hou, X. Genome-wide identification of SSR and SNP markers from the non-heading Chinese cabbage for comparative genomic analyses. *Bmc Genom.* **2015**, *16*, 328. [CrossRef] [PubMed]
2. Traw, M.B.; Bergelson, J. Interactive effects of jasmonic acid, salicylic acid, and gibberellin on induction of trichomes in Arabidopsis. *Plant Physiol.* **2003**, *133*, 1367–1375. [CrossRef] [PubMed]
3. Yang, S.; Cai, Y.; Liu, X.; Dong, M.; Zhang, Y.; Chen, S.; Zhang, W.; Li, Y.; Tang, M.; Zhai, X. A CsMYB6-CsTRY module regulates fruit trichome initiation in cucumber. *J. Exp. Bot.* **2018**, *69*, 1887–1902. [CrossRef] [PubMed]
4. Xuan, L.; Yan, T.; Lu, L.; Zhao, X.; Wu, D.; Hua, S.; Jiang, L. Genome-wide association study reveals new genes involved in leaf trichome formation in polyploid oilseed rape (*Brassica napus* L.). *Plant Cell Environ.* **2020**, *43*, 675–691. [CrossRef] [PubMed]
5. Hung, F.; Chen, J.; Feng, Y.; Lai, Y.; Yang, S.; Wu, K. Arabidopsis JMJ29 is involved in trichome development by regulating the core trichome initiation gene GLABRA3. *Plant J.* **2020**, *103*, 1735–1743. [CrossRef]
6. Sun, W.; Gao, D.; Xiong, Y.; Tang, X.; Xiao, X.; Wang, C.; Yu, S. Hairy Leaf 6, an AP2/ERF Transcription Factor, Interacts with OsWOX3B and Regulates Trichome Formation in Rice. *Mol. Plant* **2017**, *10*, 1417–1433. [CrossRef]
7. Ma, D.; Wenhua, L.; Yang, C.; Liu, B.; Fang, L.; Wan, Q.; Bingliang, L.; Mei, G.; Wang, L.; Wang, H.; et al. Genetic basis for glandular trichome formation in cotton. *Nat. Commun.* **2016**, *7*, 10456. [CrossRef]
8. Xue, S.; Dong, M.; Liu, X.; Xu, S.; Pang, J.; Zhang, W.; Weng, Y.; Ren, H. Classification of fruit trichomes in cucumber and effects of plant hormones on type II fruit trichome development. *Planta* **2018**, *249*, 407–416. [CrossRef]
9. Chen, Y.; Su, D.; Li, J.; Ying, S.; Deng, H.; He, X.; Zhu, Y.; Li, Y.; Pirrello, J.; Bouzayen, M.; et al. Overexpression of bHLH95, a basic helix–loop–helix transcription factor family member, impacts trichome formation via regulating gibberellin biosynthesis in tomato. *J. Exp. Bot.* **2020**, *71*, 3450–3462. [CrossRef]
10. Vernoud, V.; Laigle, G.; Rozier, F.; Meeley, R.B.; Perez, P.; Rogowsky, P.M. The HD-ZIP IV transcription factor OCL4 is necessary for trichome patterning and anther development in maize. *Plant J. Cell Mol. Biol.* **2009**, *59*, 883–894. [CrossRef]
11. Li, F.; Zou, Z.; Yong, H.-Y.; Kitashiba, H.; Nishio, T. Nucleotide sequence variation of GLABRA1 contributing to phenotypic variation of leaf hairiness in Brassicaceae vegetables. *Theor. Appl. Genet.* **2013**, *126*, 1227–1236. [CrossRef] [PubMed]
12. Kurlovs, A.H.; Snoeck, S.; Kosterlitz, O.; Van Leeuwen, T.; Clark, R.M. Trait mapping in diverse arthropods by bulked segregant analysis. *Curr. Opin. Insect Sci.* **2019**, *36*, 57–65. [CrossRef] [PubMed]
13. Michelmore, R.W.; Kesseli, I. Identification of markers linked to disease-resistance genes by bulked segregant analysis: A rapid method to detect markers in specific genomic regions by using segregating populations. *Proc. Natl. Acad. Sci. USA* **1991**, *88*, 9828–9832. [CrossRef]
14. Liu, D.; Yang, H.; Yuan, Y.; Zhu, H.; Zhang, M.; Wei, X.; Sun, D.; Wang, X.; Yang, S.; Yang, L. Comparative Transcriptome Analysis Provides Insights into Yellow Rind Formation and Preliminary Mapping of the Clyr (Yellow Rind) Gene in Watermelon. *Front. Plant Sci.* **2020**, *11*, 192. [CrossRef]
15. Kayam, G.; Brand, Y.; Faigenboim-Doron, A.; Patil, A.; Hedvat, I.; Hovav, R. Fine-Mapping the Branching Habit Trait in Cultivated Peanut by Combining Bulked Segregant Analysis and High-Throughput Sequencing. *Front. Plant Sci.* **2017**, *8*, 467. [CrossRef] [PubMed]
16. Liu, X.; Bi, B.; Xu, X.; Li, B.; Tian, S.; Wang, J.; Zhang, H.; Wang, G.; Han, Y.; McElroy, J.S. Rapid identification of a candidate nicosulfuron sensitivity gene (Nss) in maize (*Zea mays* L.) via combining bulked segregant analysis and RNA-seq. *Theor. Appl. Genet.* **2019**, *132*, 1351–1361. [CrossRef]
17. Wu, L.; Cui, Y.; Xu, Z.; Xu, Q. Identification of Multiple Grain Shape-Related Loci in Rice Using Bulked Segregant Analysis with High-Throughput Sequencing. *Front. Plant Sci.* **2020**, *11*, 303. [CrossRef]
18. Li, R.; Hou, Z.; Gao, L.; Xiao, D.; Hou, X.; Zhang, C.; Yan, J.; Song, L. Conjunctive Analyses of BSA-Seq and BSR-Seq to Reveal the Molecular Pathway of Leafy Head Formation in Chinese Cabbage. *Plants* **2019**, *8*, 603. [CrossRef]
19. Zheng, Y.; Xu, F.; Li, Q.; Wang, G.; Liu, N.; Gong, Y.; Li, L.; Chen, Z.-H.; Xu, S. QTL Mapping Combined with Bulked Segregant Analysis Identify SNP Markers Linked to Leaf Shape Traits in *Pisum sativum* Using SLAF Sequencing. *Front. Genet.* **2018**, *9*, 9. [CrossRef]
20. Würschum, T. Mapping QTL for agronomic traits in breeding populations. *Theor. Appl. Genet.* **2012**, *125*, 201–210. [CrossRef] [PubMed]
21. Hayes, B. Overview of Statistical Methods for Genome-Wide Association Studies (GWAS). *Methods Mol. Biol.* **2013**, *1019*, 149–169. [PubMed]
22. Barnes, S.R. RFLP analysis of complex traits in crop plants. *Symp. Soc. Exp. Biol.* **1991**, *45*, 219–228.
23. Ding, B.; Mou, F.; Sun, W.; Chen, S.; Peng, F.; Bradshaw, H.D., Jr.; Yuan, Y.-W. A dominant-negative actin mutation alters corolla tube width and pollinator visitation in Mimulus lewisii. *New Phytol.* **2016**, *213*, 1936–1944. [CrossRef] [PubMed]
24. Song, J.; Li, Z.; Liu, Z.; Guo, Y.; Qiu, L.J. Next-Generation Sequencing from Bulked-Segregant Analysis Accelerates the Simultaneous Identification of Two Qualitative Genes in Soybean. *Front. Plant Sci.* **2017**, *8*, 919. [CrossRef] [PubMed]
25. Jiao, Y.; Burow, G.; Gladman, N.; Acosta-Martinez, V.; Chen, J.; Burke, J.; Ware, D.; Xin, Z. Efficient Identification of Causal Mutations through Sequencing of Bulked F2 from Two Allelic Bloomless Mutants of Sorghum bicolor. *Front. Plant Sci.* **2018**, *8*, 2267. [CrossRef]
26. Vogel, G.; LaPlant, K.E.; Mazourek, M.; Gore, M.A.; Smart, C.D. A combined BSA-Seq and linkage mapping approach identifies genomic regions associated with Phytophthora root and crown rot resistance in squash. *TAG. Theor. Appl. Genet. Theor. Angew. Genet.* **2021**, 1–17.

27. Liu, S.; Yeh, C.T.; Tang, H.M.; Nettleton, D.; Schnable, P.S. Gene Mapping via Bulked Segregant RNA-Seq (BSR-Seq). *PLoS ONE* **2012**, *7*, e36406. [CrossRef]
28. Cheng, F.; Liu, S.; Wu, J.; Fang, L.; Sun, S.; Liu, B.; Li, P.; Hua, W.; Wang, X. BRAD, the genetics and genomics database for Brassica plants. *BMC Plant Biol.* **2011**, *11*, 136. [CrossRef]
29. Nakamura, T.; Sanokawa, R.; Sasaki, Y.F.; Ayusawa, D.; Oishi, M.; Mori, N. Cyclin I: A New Cyclin Encoded by a Gene Isolated from Human Brain. *Exp. Cell Res.* **1995**, *221*, 534–542. [CrossRef]
30. Pines, Cyclins and cyclin-dependent kinases: A biochemical view. *Biochem. J.* **1995**, *308*, 697–711. [CrossRef]
31. Wang, G.; Kong, H.; Sun, Y.; Zhang, X.; Zhang, W.; Altman, N.; Depamphilis, C.W.; Ma, H. Genome-Wide Analysis of the Cyclin Family in Arabidopsis and Comparative Phylogenetic Analysis of Plant Cyclin-Like Proteins. *Plant Physiol.* **2004**, *135*, 1084–1099. [CrossRef] [PubMed]
32. Cui, X.; Fan, B.; Scholz, J.; Chen, Z. Roles of Arabidopsis Cyclin-Dependent Kinase C Complexes in Cauliflower Mosaic Virus Infection, Plant Growth, and Development. *Plant Cell* **2007**, *19*, 1388–1402. [CrossRef]
33. Govindaraghavan, M.; Anglin, S.L.M.; Shen, K.-F.; Shukla, N.; De Souza, C.P.; Osmani, S.A. Identification of Interphase Functions for the NIMA Kinase Involving Microtubules and the ESCRT Pathway. *PLoS Genet.* **2014**, *10*, e1004248. [CrossRef] [PubMed]
34. Motose, H.; Takatani, S.; Ikeda, T.; Takahashi, T. NIMA-related kinases regulate directional cell growth and organ development through microtubule function in Arabidopsis thaliana. *Plant Signal. Behav.* **2012**, *7*, 1552–1555. [CrossRef] [PubMed]
35. Takatani, S.; Otani, K.; Kanazawa, M.; Takahashi, T.; Motose, H. Structure, function, and evolution of plant NIMA-related kinases: Implication for phosphorylation-dependent microtubule regulation. *J. Plant Res.* **2015**, *128*, 875–891. [CrossRef] [PubMed]
36. Takatani, S.; Ozawa, S.; Yagi, N.; Hotta, T.; Hashimoto, T.; Takahashi, Y.; Takahashi, T.; Motose, H. Directional cell expansion requires NIMA-related kinase 6 (NEK6)-mediated cortical microtubule destabilization. *Sci. Rep.* **2017**, *7*, 7826. [CrossRef] [PubMed]
37. Sakai, T.; Van Der Honing, H.; Nishioka, M.; Uehara, Y.; Takahashi, M.; Fujisawa, N.; Saji, K.; Seki, M.; Shinozaki, K.; Jones, M.A.; et al. Armadillo repeat-containing kinesins and a NIMA-related kinase are required for epidermal-cell morphogenesis in Arabidopsis. *Plant J.* **2007**, *53*, 157–171. [CrossRef]
38. Takatani, S.; Verger, S.; Okamoto, T.; Takahashi, T.; Hamant, O.; Motose, H. Microtubule Response to Tensile Stress Is Curbed by NEK6 to Buffer Growth Variation in the Arabidopsis Hypocotyl. *Curr. Biol.* **2020**, *30*, 1491–1503.e2. [CrossRef]
39. Saedler, R.; Jakoby, M.; Marin, B.; Galiana-Jaime, E.; Hulskamp, M. The cell morphogenesis gene SPIRRIG in Arabidopsis encodes a WD/BEACH domain protein. *Plant J. Cell Mol. Biol.* **2009**, *59*, 612–621. [CrossRef]
40. Stephan, L.; Jakoby, M.; Hülskamp, M. Evolutionary Comparison of the Developmental/Physiological Phenotype and the Molecular Behavior of SPIRRIG Between Arabidopsis thaliana and Arabis alpina. *Front. Plant Sci.* **2021**, *11*, 596065. [CrossRef] [PubMed]
41. Bögre, L.; Magyar, Z.; López-Juez, E. New clues to organ size control in plants. *Genome Biol.* **2008**, *9*, 226. [CrossRef]
42. Ren, R.; Xu, J.; Zhang, M.; Liu, G.; Yao, X.; Zhu, L.; Hou, Q. Identification and Molecular Mapping of a Gummy Stem Blight Resistance Gene in Wild Watermelon (*Citrullus amarus*) Germplasm PI 189225. *Plant Dis.* **2020**, *104*, 16–24. [CrossRef]
43. Li, H.; Durbin, R. Fast and accurate short read alignment with Burrows-Wheeler transform. *Bioinformatics* **2009**, *25*, 1754–1760. [CrossRef] [PubMed]
44. McKenna, A.; Hanna, M.; Banks, E.; Sivachenko, A.; Cibulskis, K.; Kernytsky, A.; Garimella, K.; Altshuler, D.; Gabriel, S.B.; Daly, M.J.; et al. The Genome Analysis Toolkit: A MapReduce framework for analyzing next-generation DNA sequencing data. *Genome Res.* **2010**, *20*, 1297–1303. [CrossRef] [PubMed]
45. Li, H.; Handsaker, B.; Wysoker, A.; Fennell, T.; Ruan, J.; Homer, N.; Marth, G.; Abecasis, G.; Durbin, R.; Genome Project Data Processing, S. The Sequence Alignment/Map format and SAMtools. *Bioinformatics* **2009**, *25*, 2078–2079. [CrossRef]
46. Cingolani, P.; Platts, A.; Wang, L.L.; Coon, M.; Nguyen, T.; Wang, L.; Land, S.J.; Lu, X.; Ruden, D.M. A program for annotating and predicting the effects of single nucleotide polymorphisms, SnpEff: SNPs in the genome of Drosophila melanogaster strain w1118; iso-2; iso-3. *Fly* **2012**, *6*, 80–92. [CrossRef] [PubMed]
47. Fuchu, H. E Integrated nr Database in Protein Annotation System and Its Localization. *Comput. Eng.* **2006**, *32*, 71–72.
48. Ashburner, M.; Ball, C.A.; Blake, J.A.; Botstein, D.; Butler, H.; Cherry, J.M.; Davis, A.P.; Dolinski, K.; Dwight, S.S.; Eppig, J.T.; et al. Gene Ontology: Tool for the unification of biology. *Nat. Genet.* **2000**, *25*, 25–29. [CrossRef]
49. Tatusov, R.L.; Galperin, M.Y.; Natale, D.A.; Koonin, E.V. The COG database: A tool for genome-scale analysis of protein functions and evolution. *Nucleic Acids Res.* **2000**, *28*, 33–36. [CrossRef]
50. Kanehisa, M.; Goto, S.; Kawashima, S.; Okuno, Y.; Hattori, M. The KEGG resource for deciphering the genome. *Nucleic Acids Res.* **2004**, *32*, 277–280. [CrossRef]
51. Lv, S.; Zhang, C.; Tang, J.; Li, Y.; Wang, Z.; Jiang, D.; Hou, X. Genome-wide analysis and identification of TIR-NBS-LRR genes in Chinese cabbage (*Brassica rapa* ssp. *pekinensis*) reveal expression patterns to TuMV infection. *Physiol. Mol. Plant Pathol.* **2015**, *90*, 89–97. [CrossRef]

Article

Comparative Transcriptomic Analysis of Gene Expression Inheritance Patterns Associated with Cabbage Head Heterosis

Shengjuan Li [1], Charitha P. A. Jayasinghege [2,†], Jia Guo [1], Enhui Zhang [1], Xingli Wang [1] and Zhongmin Xu [1,*]

1. College of Horticulture, Northwest A&F University, Yangling 712100, China; li2016050305@nwafu.edu.cn (S.L.); berylgj8353@nwsuaf.edu.cn (J.G.); 2008117366@nwafu.edu.cn (E.Z.); wangxingli@nwafu.edu.cn (X.W.)
2. Department of Agricultural, Food and Nutritional Science, University of Alberta, Edmonton, AB T6G 2P5, Canada; charitha.jayasinghege@canada.ca
* Correspondence: xnxzm@nwafu.edu.cn
† Present Address: Agassiz Research and Development Centre, Agriculture and Agri-Food Canada, Agassiz, BC V0M 1A2, Canada.

Citation: Li, S.; Jayasinghege, C.P.A.; Guo, J.; Zhang, E.; Wang, X.; Xu, Z. Comparative Transcriptomic Analysis of Gene Expression Inheritance Patterns Associated with Cabbage Head Heterosis. *Plants* 2021, 10, 275. https://doi.org/10.3390/plants10020275

Academic Editor: Ryo Fujimoto
Received: 23 November 2020
Accepted: 27 January 2021
Published: 31 January 2021

Publisher's Note: MDPI stays neutral with regard to jurisdictional claims in published maps and institutional affiliations.

Copyright: © 2021 by the authors. Licensee MDPI, Basel, Switzerland. This article is an open access article distributed under the terms and conditions of the Creative Commons Attribution (CC BY) license (https://creativecommons.org/licenses/by/4.0/).

Abstract: The molecular mechanism of heterosis or hybrid vigor, where F1 hybrids of genetically diverse parents show superior traits compared to their parents, is not well understood. Here, we studied the molecular regulation of heterosis in four F1 cabbage hybrids that showed heterosis for several horticultural traits, including head size and weight. To examine the molecular mechanisms, we performed a global transcriptome profiling in the hybrids and their parents by RNA sequencing. The proportion of genetic variations detected as single nucleotide polymorphisms and small insertion–deletions as well as the numbers of differentially expressed genes indicated a larger role of the female parent than the male parent in the genetic divergence of the hybrids. More than 86% of hybrid gene expressions were non-additive. More than 81% of the genes showing divergent expressions showed dominant inheritance, and more than 56% of these exhibited maternal expression dominance. Gene expression regulation by cis-regulatory mechanisms appears to mediate most of the gene expression divergence in the hybrids; however, trans-regulatory factors appear to have a higher effect compared to cis-regulatory factors on parental expression divergence. These observations bring new insights into the molecular mechanisms of heterosis during the cabbage head development.

Keywords: *Brassica oleracea* L. var. capitata; heterosis; transcriptomics; allele-specific expression; cis-and trans-regulation

1. Introduction

The phenomenon where offspring of genetically diverse parents show superior or beneficial alterations of agronomic traits, such as growth potentials, yield, fertility, or stress tolerance, compared to their parents, is known as heterosis or hybrid vigor [1]. Plant breeding programs widely benefit from this phenomenon in obtaining desired or improved crop qualities [2]. Cabbage (*Brassica oleracea* L. var. *capitata*) is one of the most consumed leafy vegetables in the world [3]. Since most of the commercial cultivars of cabbage are F1 hybrids obtained by outcrossing, understanding the underlying mechanisms of heterosis can help increase cabbage breeding efficiencies and improve the market qualities of cabbage, such as head size and appearance.

Two classical genetic models, "dominance" and "overdominance", have been widely discussed as potential genetic effects of heterosis [1,4]. According to the dominance model, complementation of multiple deleterious recessive alleles of each parent by superior, dominant alleles of the other parent results in heterosis. The overdominance model describes heterosis as a result of synergistic interactions between alleles at heterozygous loci of a hybrid. Nevertheless, these models can only explain single-locus heterosis [5]. Since most of the traits correlated to heterosis are controlled by multigenic effects, perspectives of energy-use efficiency and protein metabolism are commonly used to explain general multigenic heterosis [6–8]. Also, transcriptomic analysis has become a valuable tool that can be used to broaden our understanding of heterosis. For example, differentially expressed genes (DEGs) identified by transcriptomics can be used to predict crucial physiological pathways [9]. The data can also be used to identify genetic variations, including single nucleotide polymorphisms (SNPs) and small insertions/deletions (InDels) as well as imbalanced contributions of the parental genomes in the hybrid gene expression [10,11].

Gene expression divergence between hybrid and parents can be classified into multiple gene expression inheritance patterns. In additive expression inheritance, the transcript level of a given gene in a hybrid is similar to mid-parent. In non-additive inheritance, the hybrid transcript level can be similar to the parent showing the highest or lowest expression (high or low parent dominance; also referred to as positive or negative non-additivity), or it can be higher or lower than both parents (transgressive-up and transgressive-down; also referred to as overdominant and underdominant expression levels) [12–14]. All these expression inheritance patterns may be found in a single hybrid and can contribute to heterosis in different proportions [9]. For example, an expressed sequence tag (EST) microarray analysis in the F1 hybrids of maize showed the presence of all possible expression patterns; however, 78% of the differentially expressed ESTs exhibited additive expressions [12]. In *Nicotiana tabacum* L. hybrids, paternal expression level dominance was the most prominent, but 13 key genes regulating nicotine anabolism and transport showed transgressive-up or -down regulations [15].

Allele-specific regulation of gene expressions during environmental and stress responses indicates that allelic variations may play a fundamental role in the heterosis [16]. Allelic differences are widespread in hybrids and can have significant effects on gene expressions [17]. Parental alleles interacting with each other can result in cis- and trans-regulations. The cis-effects, such as the changes in promoters and enhancers, affect gene expression in a single chromosome and therefore are restricted to a chromosome from one parent. In contrast, the effects of trans-regulations, such as the changes in transcription factors, can affect chromosomes from both parents [18]. Even though both cis- and trans-regulations contribute to divergent gene expressions, their respective contributions are not clear. For example, in maize, allelic cis-regulatory variation between some inbred lines largely contributes to the gene expression divergence in the F1 hybrids [19]. In *Cirsium arvense*, the trans-effect shows a greater correlation with the parental expression divergence and tends to drive the higher expressions of paternal alleles [20]. In the F1 hybrids between *Arabidopsis thaliana* and *Arabidopsis arenosa*, both cis- and trans-regulations appear to mediate gene-expression divergence and chromatin modifications [21].

While heterosis is essential in cabbage breeding, previous studies of cabbage have mostly focused on phenotypes over underlying mechanisms [22,23]. Among the few studies evaluating genetic mechanisms, several have examined quantitative trait loci associated with the head shape or other quality-related traits [24,25]. Microarray-based analysis has also been used to search for potential regulatory genes related to heterosis [26]. However, none of these studies reflect the comprehensive gene expression changes between cabbage hybrids and their parents. In this study, we examined horticultural traits and transcriptome profiles of four F1 cabbage hybrids and their parents. We evaluated the genetic and gene expression divergences between hybrids and parents to understand the genetic regulatory mechanisms and their contribution to heterosis. Cellular processes represented by DEGs, transcription factors associated with cabbage head growth and development, and the cis- and trans-effects were investigated. Our research provides a comparative perspective on the gene expression inheritance patterns between cabbage hybrids and their parents.

2. Results

2.1. Cabbage Hybrids Show Heterosis in Several Horticultural Traits

Three lines of male parents, QP03, DHP37, and QP15 (hereafter referred to as MP1, MP2, and MP3, respectively), three lines of female parents, QP13, QP04CMS, and QP16CMS (hereafter referred to as FP1, FP2, and FP3, respectively), and four of their F_1 hybrids, HY1 (FP1 × MP1), HY2 (FP2 × MP1), HY3 (FP2 × MP2), and HY4 (FP3 × MP3) were studied (Figure 1a). The net weight of cabbage heads, polar head diameter, equatorial head diameter, weight of the distinguishable petioles, non-wrapper leaf weight, plant height, and plant diameter were evaluated and showed significant differences in the hybrids compared to their parents (Figure 1b–h). In addition, mid-parent heterosis (MPH) and high-parent heterosis (HPH)—the percentage deviation of the hybrid means (M_{HY}) from the mean of the mid-parent (M_{MiP}) and the high-parent (M_{HiP}), respectively—were calculated. The MPH values were positive for all the traits in all the hybrids, which demonstrates improvements compared to the average of their parents. The HPH values, except for the non-wrapper leaf weight and plant height in HY1 and equatorial head diameter in HY3, were also positive, indicating improvements over the highest of the two parents. The increase in cabbage head weight was the most prominent, with MPH and HPH values ranging from 107–129% and 87–100%, respectively (Table 1), which suggests higher yield potentials of the hybrids.

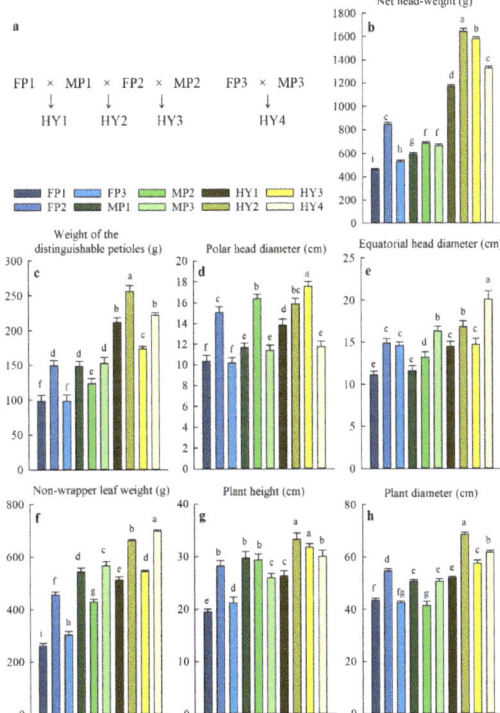

Figure 1. The crosses used for the generation of hybrids and the differences in the cabbage head and plant size traits. (**a**) The female parent (FP) and the male parent (MP) crosses used to obtain different cabbage hybrids (HY). HY1 and HY2 share the MP; HY2 and HY3 share the FP; HY4 does not share parents with any other hybrid. (**b–h**) Variation of cabbage head size and plant size among cabbage hybrids and parents. Data are means ± SD, n = 3 blocks, with each block representing the average of 10 cabbages. Different letters denote statistical differences as determined by one-way ANOVA coupled with LSD post-hoc test ($p < 0.05$).

Table 1. Mid-parent heterosis (MPH) and high-parent heterosis (HPH) values for different horticultural traits of the cabbage hybrids.

Hybrid	Net Head-Weight	Weight of the Distinguishable Petioles	Polar Head Diameter	Equatorial Head Diameter	Non-Wrapper Leaf Weight	Plant Height	Plant Diameter
MPH (%) ‡							
HY1	121.9 ± 2.2	71.7 ± 4.9	25.9 ± 0.6	27.9 ± 0.7	27.3 ± 1.0	7.1 ± 1.4	10.2 ± 0.7
HY2	129.2 ± 4.4	72.6 ± 6.8	18.9 ± 0.6	27.4 ± 0.7	32.4 ± 2.7	15.1 ± 1.3	30.2 ± 0.2
HY3	106.7 ± 3.5	27.1 ± 4.8	11.8 ± 0.7	5.0 ± 0.8	23.3 ± 2.3	10.4 ± 1.8	19.8 ± 0.3
HY4	122.6 ± 4.1	77.4 ± 10.0	9.1 ± 0.2	30.3 ± 2.2	60.5 ± 4.0	27.5 ± 0.6	32.3 ± 0.6
HPH (%)							
HY1	97.5 ± 2.3	42.6 ± 2.0	18.6 ± 0.7	25.1 ± 1.3	−5.9 ± 0.2	−11.3 ± 1.3	2.3 ± 0.3
HY2	94.5 ± 3.2	72.1 ± 6.8	5.4 ± 0.7	13.2 ± 0.6	21.7 ± 2.4	12.3 ± 1.8	25.5 ± 0.5
HY3	86.7 ± 3.1	16.1 ± 3.9	7.1 ± 0.3	−0.9 ± 1.1	19.6 ± 2.2	7.9 ± 5.1	5.3 ± 0.8
HY4	99.9 ± 3.2	45.9 ± 6.2	3.1 ± 0.3	23.3 ± 2.1	23.1 ± 2.3	15.8 ± 0.5	21.7 ± 0.9

‡ The percentage deviation of the hybrid means from the mean of the mid-parent and the high-parent. Data are means ± SD (n = 3 blocks, with each block representing the average of 10 cabbages).

2.2. Genetic Variations and Gene Expression Divergence between Hybrids and Their Parents

Transcriptome sequencing of the four cabbage hybrids and their six parental lines (each with three biological replicates) generated a total of 230.65 Gb of high-quality clean reads with a minimum of 6.28 Gb clean data for each replicate. A total of 771.2 million paired-end clean reads were sequenced with a Q30 score of ≥90.9%. The RNA-seq data was deposited to the NCBI sequence read archive (SRA) under the accession number PRJNA664256. After filtering out low quality reads, between 42.0 and 61.8 million paired-end reads were obtained for each replicate. Of these, about 67–70% of the paired reads aligned with the *B. oleracea* var. *capitata* reference genome and 66–68% of them aligned to unique positions. The number of reads aligned with positive and negative chains was almost identical.

Genome-wide genetic variations play a substantial role in heterosis [10]. To determine the frequency of shared genetic variations, we examined the SNPs and InDels in the hybrids and parents. Compared to the reference genome, HY1–HY4 had 160,928, 177,148, 180,019, and 172,891 SNPs and 9112, 9617, 9743, and 9353 InDels, respectively. We observed that 3–12% of SNPs and 5–10% of InDels were unique to the hybrids. Also, 26–28% of SNPs and 14–15% of InDels were detected in the genomes of the hybrids and their two parents, with reference to the *B. oleracea* var. *capitata* reference genome. When the remaining SNPs and InDels were considered, the number of SNPs and InDels consistent with the female parent were slightly but consistently higher than that of the male parent, indicating a maternal parent bias in the transcriptome. This bias was particularly prominent in HY1, with 75,299 SNPs and 2359 InDels consistent with the female parent, compared to only 54,757 SNPs and 1642 InDels consistent with the male parent (Table 2).

To evaluate the hybrid gene expression deviations from their parents, we performed pairwise expression comparisons and identified DEGs with more than two-fold expression difference and a false discovery rate of ≤0.01. Interestingly, all hybrids had a higher number of DEGs when compared with the male parent than when compared with the female parent (Table 3). This lower expression divergence between hybrids and female parents, together with the allelic expression bias towards female parents, suggests a closer genetic association between hybrids and female parents than hybrids and male parents.

Table 2. The genetic variations between cabbage hybrids and their parents.

Genetic Variation	Hybrid			
	HY1	HY2	HY3	HY4
SNPs				
Number of hybrid specific SNPs	5180	11,057	22,290	9698
Number of SNPs consistent with both parents	44,795	47,246	46,570	45,092
Number of SNPs consistent with the male parent	54,757	72,233	62,763	72,271
Number of SNPs consistent with the female parent	75,299	77,013	76,463	72,426
InDels				
Number of hybrid specific InDels	457	677	921	591
Number of InDels consistent with both parents	1309	1415	1433	1380
Number of InDels consistent with the male parent	1642	2124	2007	2189
Number of InDels consistent with the female parent	2359	2428	2398	2258

Table 3. Comparisons of DEGs between hybrids and parents, or among hybrids or parents.

Comparison Group	Number of DEGs	Up-Regulated	Down-Regulated
FP1 × MP1→HY1			
FP1 vs. HY1	48	25	23
MP1 vs. HY1	3191	1562	1629
FP2 × MP1→HY2			
FP2 vs. HY2	1773	1142	631
MP1 vs. HY2	2053	1272	781
FP2 × MP2→HY3			
FP2 vs. HY3	2170	1454	716
MP2 vs. HY3	3420	2098	1322
FP3 × MP3→HY4			
FP3 vs. HY4	1334	874	460
MP3 vs. HY4	1952	1118	834
Hybrids/parents			
HY1 vs. HY2	948	523	425
HY1 vs. HY3	1294	870	424
HY1 vs. HY4	1991	1082	909
HY2 vs. HY3	2342	1293	1049
HY2 vs. HY4	2872	1208	1664
HY3 vs. HY4	2475	1053	1422
MP1 vs. MP2	5139	2369	2770
MP1 vs. MP3	4391	2047	2344
MP2 vs. MP3	4915	2503	2412
FP1 vs. FP2	2011	1132	879
FP1 vs. FP3	3683	1933	1750
FP2 vs. FP3	3699	1792	1907

In our hybrids, HY1 and HY2 shared the same male parent (MP1), and HY2 and HY3 shared the same female parent (FP2; Figure 1a). The lower gene expression divergence between hybrids and the female parent suggests that HY2 and HY3 may have a lower expression divergence compared to HY1 and HY2 (Table 3). However, there were 2342 DEGs between HY2 and HY3 compared to only 948 DEGs between HY1 and HY2 (Table 3). This discrepancy is likely due to the higher number of DEGs between MP1 and MP2 (5139) compared to that between FP1 and FP2 (2011) (Table 3).

We further evaluated the DEGs between hybrids and parents as well as between male and female parents using Venn diagrams to identify common DEGs among hybrids (Figure 2). When hybrids and their male parents were compared, there were 145 DEGs common to all hybrids (Figure 2a). The comparison between hybrids and their female parents, however, revealed only two common DEGs (Figure 2b). This lower number of DEGs common to hybrids and their female parents can be due to the relatively smaller number of DEGs between hybrids and female parents; nevertheless, it also suggests a higher and steady divergence of hybrids from male parents.

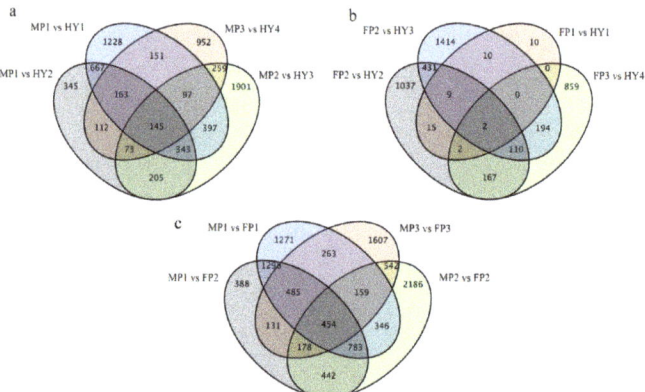

Figure 2. Venn diagram comparison of differentially expressed genes (DEGs) between hybrids and their parents. The common DEGs in different hybrids compared to their (**a**) male parent and (**b**) female parent, and (**c**) between female and male parent combinations used to generate the four different hybrids.

2.3. Functions of DEGs Associated with Cabbage Head Heterosis

To determine the roles of DEGs, we performed GO (Gene Ontology), COG (Clusters of Orthologous Groups), and KOG (Eukaryotic Orthologous Groups) analysis, as these databases cover detailed gene or protein function information. Among the four hybrids, HY3 had the highest number of DEGs compared to its parents. As the higher number of DEGs suggests stronger divergence of hybrids from their parents, GO, COG, and KOG pathway enrichment analysis between HY3 and its parents, MP2 and FP2 (MP2 vs. HY3 and FP2 vs. HY3), are described below; however, the evaluation of DEGs between other hybrids and their parents produced similar results.

In the GO enrichment analysis, "cellular processes" was the most overrepresented biological process subcategory, with 1479 and 2237 DEGs in HY3 compared to MP2 and FP2, respectively (the DEGs are given in the same order in all the following enrichment categories). "Single-organism process" (1411 and 2120) and "metabolic process" (1360 and 2053) were the other most represented subcategories. In the cellular component and molecular function categories, "cell part" (1738 and 2723) and "binding" (1060 and 1632) subcategories were the most enriched terms, respectively (Figure S1). In COG analysis, "transcription" (115 and 187), "signal transduction mechanisms" (117 and 164), and "replication, recombination, and repair" (107 and 161) were discovered as the top three enriched terms (Figure S2). The most enriched categories in KOG analysis were "posttranslational modification, protein turnover, chaperones" (102 and 191), "signal transduction mechanisms" (102 and 177), and "carbohydrate transport and metabolism" (94 and 127; Figure S2).

To evaluate the transcription factors that may play a role in heterosis, we annotated the transcription factors represented by DEGs using the plant transcription factor database (PlantTFDB v4.0). A total of 54 MYB or MYB-related, LATERAL ORGAN BOUNDARIES (LOB) domain, and the KNOTTED1-LIKE HOMEOBOX (KNOX) transcription factors are among the differentially expressed genes that are likely involved in the growth and development of cabbage head leaves. The KNOX, MYB, and LOB domain transcription factors interact with each other, regulating the shoot morphogenesis and leaf patterning in the apical meristem [27,28]. The BELL-like homeobox and HOMEO-DOMAIN LEUCINE ZIPPER (HD-ZIP) transcription factors, which establish the polarity and leaf outgrowth [29], and the AUXIN RESPONSE FACTORs, which regulate the expression of auxin-responsive genes [30], were also abundant (Table S1). These diverse categories indicate the regulatory roles of extensive biosynthetic, metabolic, and signal transduction pathways behind heterosis.

2.4. Gene Expression Inheritance Patterns in the Cabbage Hybrids

To gain an overall insight into gene expression changes, we classified the inheritance patterns of DEGs as additive and non-additive. The non-additive inheritance was divided into paternal expression dominance, maternal expression dominance, and transgressive expressions. The transgressive category was further divided to differentiate up and down regulations, and the parental dominance categories were further divided to differentiate high-parent dominance and low-parent dominance. In total, there were eight different expression inheritance categories (I–VIII; Figure 3). Approximately 66–70% of the DEGs in the hybrids showed an expression level dominance, 9–14% showed an additive inheritance pattern, and 17–25% were transgressive. Among those exhibiting dominance, the number of genes showing high-parent dominance was higher than those showing low-parent dominance (Figure 3; compare category III with IV, and V with VI). Similarly, transgressive up-regulation was predominant over transgressive down-regulation (Figure 3; compare category VII with VIII). Of the genes showing expression level dominance, 53–74% showed maternal expression level dominance, whereas only 26–47% showed paternal expression level dominance, which again indicates a maternal bias in the expression inheritance (Figure 3; categories III–VI).

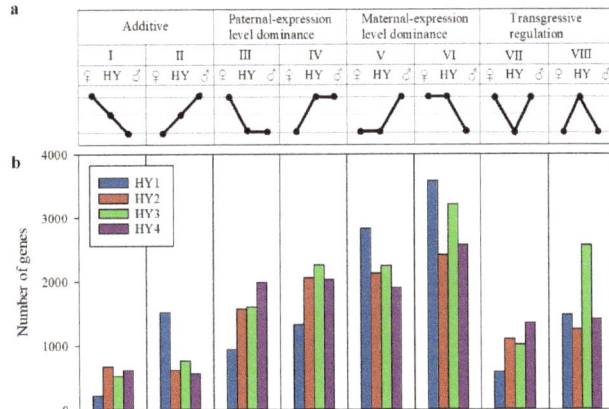

Figure 3. Gene expression inheritance patterns in the cabbage hybrids. (**a**) The eight different gene expression inheritance patterns based on the hybrid gene expression level compared to its parents: additive (I and II), paternal expression level dominance (low-parent dominance and high-parent dominance; III and IV), maternal expression level dominance (low-parent dominance and high-parent dominance; V and VI), transgressive regulation (upregulation and downregulation; VII and VIII). (**b**) The number of genes representing each expression inheritance category in the four hybrids.

To further evaluate the maternal expression bias, we selected the genes with parent-specific SNPs and determined relative allele-specific expressions (ASEs), the percentage of female parent alleles in the transcriptome (% FP_{HY}). In each hybrid, representation of female parent alleles in the transcriptome for approximately 41–48% of the genes was between 40–60% (40–60 category in Figure 4a), and therefore, these genes showed no clear expression bias [31]. For the remaining genes, HY1 and HY3 showed a maternal bias (60–100% category), but the other two hybrids showed no clear bias to either parent (Figure 4a). We further plotted the expression ratio of each allele in the parents (FP/MP) against their expression in the hybrids (FP_{HY}/MP_{HY}) on a logarithmic scale (Figure 4b).

The distributions of alleles in these plots were mostly symmetrical and showed no clear bias to either parent. However, in HY1, more gene position in the upper quadrants indicates higher expressions of FP alleles (Figure 4b).

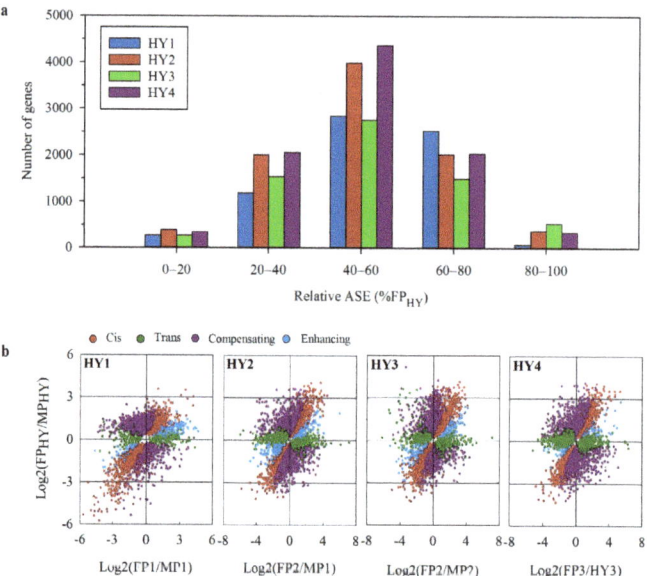

Figure 4. Relative allele-specific expression in the cabbage hybrids. (**a**) The number of genes representing differential expressions as grouped by the relative expression level of maternal alleles (% FP_{HY}). Genes in the 40–60 category were considered to have a balanced allelic expression. Genes in the 60–80 and 80–100 categories show a female allele bias. Genes in the 0–20 and 20–40 categories show an expression bias toward the male allele. (**b**) The log2 expression ratios of maternal to paternal alleles in the parents vs. hybrids. Each point represents a single gene with colors representing the regulatory divergence category. MP_{HY} and FP_{HY}: maternal and paternal allelic expression levels in the hybrids, respectively; MP and FP: maternal and paternal expression levels in the parents, respectively.

In all the hybrids, regardless of the expression inheritance patterns, the distribution ratio of relative ASEs was always unbalanced (Figure 5; see Figures S3–S5 for HY2–HY4). In the paternal expression dominance category showing high-parent dominance (category IV), relative ASE showed a paternal bias (0–40% category is predominant). In the maternal expression dominance category showing high-parent dominance (category VI), relative ASE also showed a maternal allele bias (60–100% category is predominant). In contrast, in the categories showing low-parent paternal and maternal dominance (categories III and V, respectively), relative ASE showed maternal and paternal allele bias, respectively. In the categories showing additive inheritance patterns, a maternal allele bias was observed in additive female parent > male parent conditions (category I), and a paternal allele bias was observed in additive male parent > female parent conditions (category II). The only exception was for HY1, where a paternal allele bias was observed in both situations. As for the transgressive up-regulation and down-regulation categories (categories VII and VIII), HY1 and HY2 showed a maternal allele bias, but no clear association could be seen in the other hybrids (Figure 5 and Figures S3–S5). The results suggest that relative ASE always shows a bias towards the parent showing the higher expression.

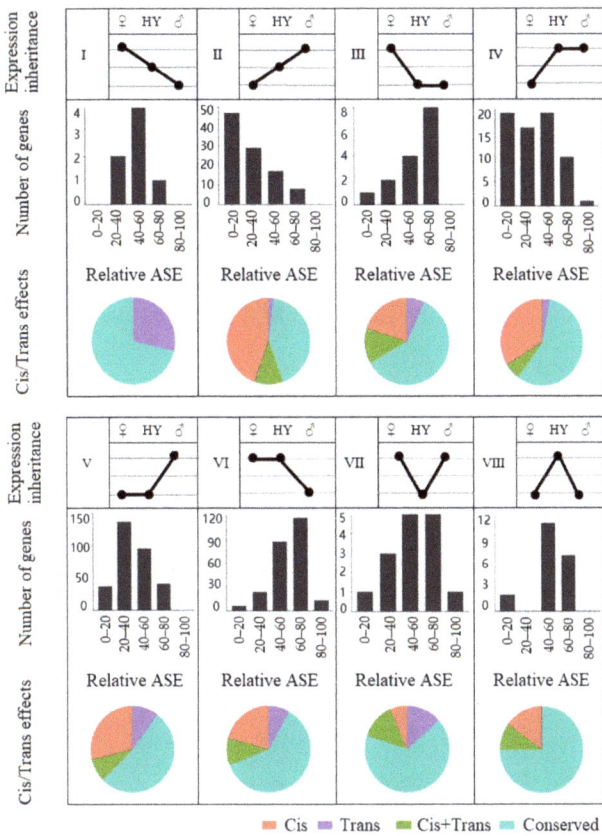

Figure 5. Comparison of relative allele-specific expressions (ASEs) and cis- and trans-effects according to seven expression inheritance patterns in HY1. The relative ASE represents the expression of maternal alleles as a percentage of the total gene expression in the hybrid (% FP_{HY}). The pie charts show the proportion of genes showing cis-effects, trans-effects, or both cis- and trans- (cis + trans) effects. Genes showing no significant evidence of cis- or trans-effects were classified as conserved.

2.5. Cis- and Trans-Effects on Gene Expression Divergence

The gene expression divergence between hybrids and parents can result from both cis- and trans-regulatory changes [32]. To explore the genetic basis of expression divergence, we used parent-specific SNPs to compare ASE. After applying the quality control criteria, 31,962 (7223 genes), 52,177 (9246 genes), 34,653 (6937 genes), and 56,099 (9628 genes) SNPs with ≥10 read coverage were screened in HY1–HY4, respectively. Among them, 45–56% of the genes showed no expression divergence in the hybrids compared to their parents and were classified as conserved expressions (Figure 6a). The alleles that expressed differently between the parents and maintained that differential expression in the hybrids were classified as having only cis-effects. The alleles that expressed differentially in the parents but expressed equally in the hybrid were considered to have only trans-effects. In the remaining genes, cis- and trans-effects were co-acting (cis + trans effects) [11]. Cis-effects accounted for the majority of expression divergence in all the hybrids except HY3, where compensating cis + trans effects were predominant. Among the four hybrids, 21–26% of the genes analyzed showed cis effects, and 13–16% showed trans-effects (Figure 6a). The compensating cis + trans effects, where cis- and trans-effects act in the opposite directions, accounted for 13–23% of the ASE. In contrast, only 1–2% of the ASE was represented by enhancing cis + trans interactions, where cis- and trans-effects act in the same direction (Figure 6a).

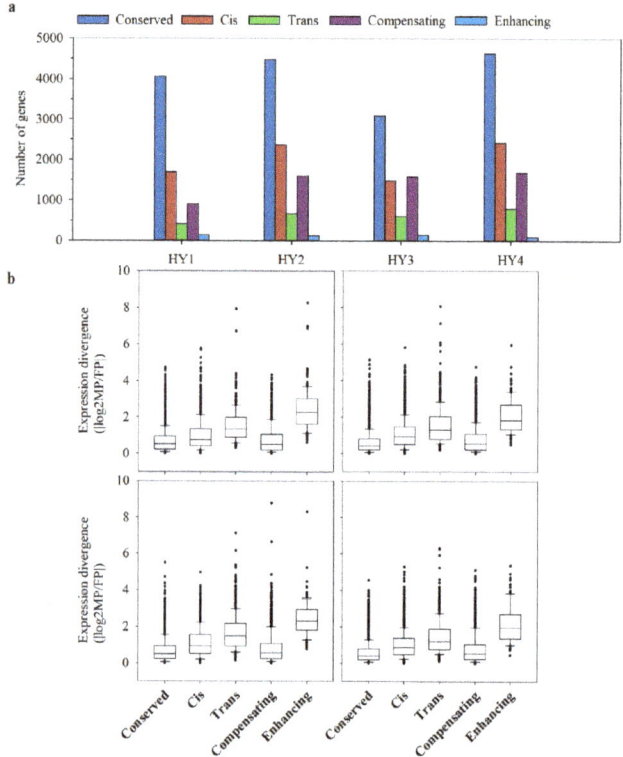

Figure 6. Cis- and trans-effects in the hybrids. (**a**) The number of genes in different regulatory categories. The "conserved" category represents the genes showing no clear cis- or trans-effects. Genes showing evidence of both cis- and trans-effects were subdivided as "compensating" and "enhancing", where the cis- and trans-effects act in opposite directions and the same direction, respectively. (**b**) The absolute magnitude (fold-change) of parental expression divergence resulting from different regulatory effects. The median trans-effects were larger than the median cis-effects in all the four hybrids.

The contribution of cis- or trans-effects to the gene expression differences of the parents was inspected from the absolute magnitude of parental expression divergence in the hybrids. First, the median of both cis- and trans-regulations showed a high level of expression divergence relative to the conserved gene expression in all the hybrids, which indicates that both cis- and trans-regulations are behind parental expression divergence (Figure 6b). Second, although there are more transcripts with only cis-effects than with only trans-effects, the trans-effect appeared to contribute more to the gene expression divergence than the cis-effect (Figure 6b).

2.6. The Relationship between Allelic Expression Regulation and Expression Inheritance in Hybrids

Trans-regulatory factors commonly show asymmetric effects on the expression of parental alleles in hybrids, especially in allopolyploids [31]. For example, in F1 allotetraploids between *A. thaliana* and *A. arenosa*, the *A. arenosa* trans-factors tend to upregulate *A. thaliana* alleles, whereas *A. thaliana* trans-factors either upregulate or downregulate

A. arenosa alleles [21]. To determine the roles of trans-regulatory factors in the parental expression bias of cabbage hybrids, we further assessed ASE variations in our hybrids. As the cis-regulatory factors in a parent genome and the hybrid genome representing that parent are similar, expression variations of alleles represent the effects of trans-regulatory factors [31]. To visualize the effects of trans-regulatory factors on the gene expression, we plotted the relative ASE of male parent alleles (MP_{HY}/MP) against that of the female parent alleles (FP_{HY}/FP) on a log2 scale [31]. The mostly symmetrical distribution of points in the plot shows that trans-regulatory factors from each parental genome can equally upregulate or downregulate alleles from the other genome, and the magnitude of expression variation in MP_{HY} alleles is similar to FP_{HY} alleles (Figure 7). Therefore, the FP and MP trans-regulatory factors appear to have a similar contribution to the hybrid gene expression divergence. The only exception was in HY1, where the expression variation of $MP1_{HY1}$ alleles was higher than that of the $FP1_{HY1}$ alleles, suggesting a higher impact of FP1 trans-regulatory factors on the $MP1_{HY1}$ alleles compared to MP1 trans-regulatory factors on $FP1_{HY1}$ alleles.

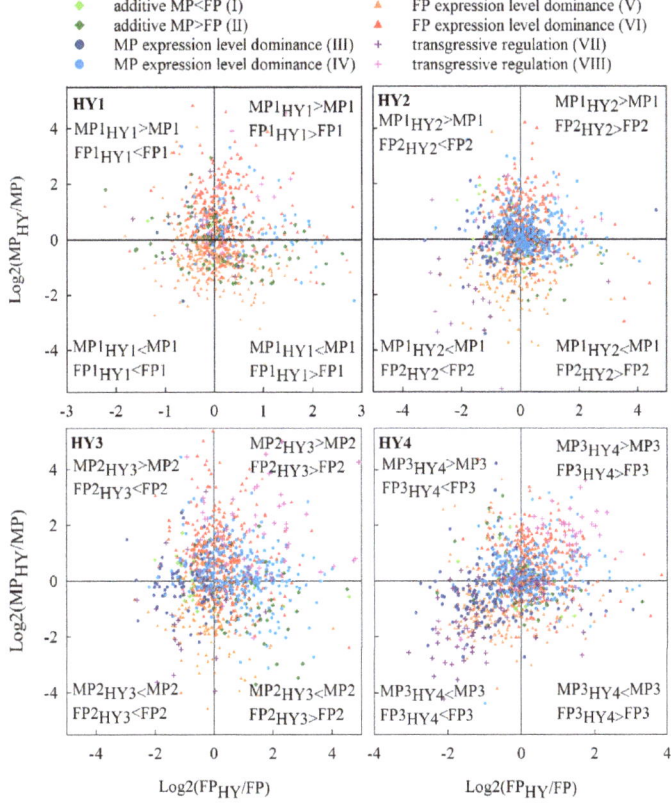

Figure 7. Allele-specific expression (ASE) variations of the differentially expressed genes (DEGs) in the hybrids. Each point in a plot represents a single gene. The position of each point represents the combined expression change of the two alleles in the hybrid with respect to the male (MP) and female (FP) parent.

The distribution pattern of the genes in the scatter plot also shows that in the majority of genes classified as showing maternal expression level dominance with a high-parent dominance (category VI in Figure 3), the MP_{HY} alleles were mainly upregulated by FP trans-regulatory factors. In the genes showing a maternal expression level dominance with a low-parent dominance (category V), the MP$_{HY}$ alleles were mostly downregulated by FP trans-regulatory factors. The same trend was observed in the paternal dominance. In the genes showing paternal expression level dominance with a high-parent dominance (category IV), the FP$_{HY}$ alleles were mainly upregulated by MP trans-regulatory factors. In the genes showing a paternal expression level dominance with a low-parent dominance (category III in Figure 3), the FP$_{HY}$ alleles were mostly downregulated by MP trans-regulatory factors. For the additive and transgressive categories, gene expression appears to be a combined effect of both male parent and female parent trans-regulatory factors (Figure 7). Overall, the expression level dominance towards a parent can be explained by the effect of that parent's trans-regulatory factors on the other parental genome.

3. Discussion

3.1. Gene Expression Divergence between Cabbage Hybrids and Their Parents

In this study, we used transcriptome sequencing to evaluate the genomic and transcriptomic level changes underlying cabbage head heterosis. Many transcriptomic analyses in plants and other organisms show that gene expression in hybrids can be biased towards a particular parent [14,20,31,33–35]. In cabbage hybrids, the higher number of SNPs and InDels consistent with the female parent than with the male parent, suggests a closer genetic relationship between the female parent and the hybrids. The number of DEGs showing expression profiles similar to the female parent was also higher compared to male parents in all the hybrids. Thus, we deduced that the female parent likely plays a larger role than the male parent in the hybrid gene expression divergence. However, the present analysis was limited to a single stage of cabbage head development, and the gene expression bias may shift depending on the developmental stage. For example, in rice (*Oryza sativa* L. ssp. *indica*), transcriptome profiles of leaves were closer to the maternal parent at the early plant development, but closer to the paternal parent at later stages [34]. Also, without reciprocal crosses, we cannot rule out other possible factors that may cause the observed female parent bias. The parent-of-origin effects, also known as transgenerational effects, are associated with the parental genotype and can be influenced by the environmental conditions or physiological state. For example, the maternal genotype of the endosperm affects seed development, which may influence early seedling development and thereby plant vigor at later stages [36]. The parent-of-origin effects may also arise due to genomic imprinting, an epigenetic regulatory mechanism that causes one parental allele to be expressed prominently. In plants, genomic imprinting is usually limited to multiple genetic loci with single genes, and imprinted genes are almost entirely confined to endosperm. Therefore, genomic imprinting generally affects reproductive development [37]; however, we cannot exclude the possibility that this may cause long-term effects that appear in plants at later stages.

The other key questions here are whether the gene expression bias towards a particular parent has a positive or negative correlation with heterosis, and whether it is a cause or consequence of heterosis. Even though the answers to these questions are not clear, in rice, analysis of some selected loci suggests a negative correlation between hybrid yield and paternal gene expression bias [35]. In maize (*Zea mays* L.), a smaller positive relationship was reported between the yield and maternal expression bias [38]. Even so, whether these associations are responsible for heterosis or are merely a consequence of the phenotypes remains unknown.

The high HPH and MPH values show that there is a significant improvement in the head weight and petiole weight of all the hybrids (Table 1). Many studies show that yield or biomass heterosis correlates with an increased level of metabolic activity [12,34,39]. In maize yield heterosis, for example, DEGs have been found to be significantly enriched in carbohydrate metabolism associated genes [40]. At the heading stage of rice, the DEGs mapped to quantitative trait loci (QTLs) for yield were also linked to carbohydrate metabolism [41]. We also observed a significant enrichment of carbohydrate transport and metabolism genes in the COG analysis of DEGs, which suggests a correlation in biomass heterosis and carbohydrate metabolism in cabbage hybrids.

In the GO function, COG, and KOG pathway enrichment analyses, metabolic process, catalytic activity, replication, recombination and repair, transcription, and signal transduction mechanisms were among the most overrepresented terms (Figures S1 and S2). Many of these terms are commonly represented in the heterosis of various other plant species. For example, evaluation of DEGs associated with rice seedling heterosis has shown that transcription, metabolism of cofactors and vitamins, amino acid metabolism, and biosynthesis pathways of secondary metabolites are significantly enriched [42]. Genome-wide comparisons of maize hybrids indicated that biological processes, including metabolism, signal transduction, transport, biological regulation, and development are the main functions of the genes represented by DEGs [43]. In Arabidopsis, pathways contributing to growth heterosis were also composed of enrichment categories similar to those observed in cabbage [44]. Altogether, the similarity of enriched pathways associated with different growth-related traits in diverse plant species suggests that those traits are mostly regulated by broader but relatively similar pathways rather than a single gene or locus affecting a specific pathway [45].

3.2. Maternal Expression Level Dominance is Predominant in Cabbage Hybrids

Both additive and non-additive gene expression inheritance patterns contribute to the gene expression divergence in hybrids; however, the relative contribution of each inheritance pattern to heterosis is not clear. In our cabbage hybrids, more than 86% of genes showed non-additive expression inheritance patterns. This high percentage indicates that the expression divergence of most hybrid genes is not merely a result of the combined effect of allelic expressions, but a consequence of diverse gene expression regulatory mechanisms. A number of recent studies, including in maize [46], soybeans [47], rubber trees [10], interspecific hybrids between *Brassica napus* and *B. rapa* [48], and cauliflower [49], show a prevalence of non-additive gene action in the hybrids. Non-additive gene expressions may have both direct and indirect contributions to heterosis. For instance, in *Arabidopsis* hybrids, non-additive genes likely enhance metabolic activities, which leads to improved resource utilization and increased seedling growth rates [50]. In soybeans, 19 non-additive genes associated with nitrogen use efficiency likely play a role in heterosis by improving the protein content of seeds [47].

The expression of a given gene in a hybrid can be non-additive when it is showing a maternal or paternal expression dominance. Also, the expression of some genes can be independent of the parental expression levels (transgressive regulation). Of the genes showing non-additive expressions in our hybrids, 72–81% (66–70% of the total genes) showed dominant expression patterns. Approximately 53–74% of these genes showed maternal expression level dominance, and only 26–47% showed paternal expression level dominance. Therefore, it is likely that cabbage hybrids predominantly show an expression level dominance with a maternal bias when establishing heterosis. This observation is consistent with various other studies, where parental expression dominance has been shown to predominate in the hybrids [15,31,48,51].

3.3. Cis-Regulation Predominates the Hybrid Gene Expression Divergence

Gene expression is controlled by a complex regulatory network, which includes interactions between DNA, RNA, proteins, and environmental factors. The quantitative changes in gene expressions, however, are directly regulated by the cis- and trans-effects [32]. If differences of allelic regulation are only due to cis-regulatory changes, the expression of the allele in hybrids can be expected to be additive. However, hybridization exposes the parental alleles to trans-regulatory factors originating from both parents, and therefore, the differences in allelic regulation may also be due to trans-effects. Numerous studies have shown that non-additive gene expression mainly results from the trans-effects, and may cause the gene expression in the hybrids to deviate significantly from additive expression [31,46,51,52]. In cabbage hybrids, of the genes showing non-conserved expressions, 39–54% showed cis-effects, compared to only 13–16% showing trans- effects (Figure 6a). However, in the parents, the majority of gene expression differences were regulated by trans-regulatory factors (Figure 6b).

Plants have to adapt to the changing environment continually. Enhancing cis + trans interactions increase gene expression divergence and promote disruptive or diversifying plant characteristics. In contrast, compensating cis + trans interactions reduce gene expression divergence (stabilizing selection), since the cis effects are compensated by opposite actions of trans-effects or vice versa. [21]. Stabilizing selection tends to keep the stability of internal cellular functions by maintaining the expression of genes involved in metabolic and biosynthetic processes at balanced levels [53]. Among the genes showing non-conserved expressions, 29–41% showed compensating cis + trans interactions, compared to only 2–4% showing enhancing cis + trans interactions (Figure 6a). Therefore, stabilizing selection effects appear to be common in maintaining the gene expression levels in cabbage hybrids. Altogether, it is likely that a complex combination of cis- and trans- effects determines the gene expression inheritance patterns.

Since trans-effects increased with the parental expression divergence, we further assessed the role of trans-effects in modulating allelic expression and gene expression inheritance patterns. The mostly symmetrical distribution of alleles in Figure 7 suggests that the effects of trans-regulatory factors from the two parents are generally balanced. The regulation of non-dominant parent alleles by the trans-regulatory factors of the dominant parent, without a bias towards a parent, likely caused the majority of the expression level dominance. The other inheritance patterns are likely determined by the combined effect of trans-regulatory factors from both parents acting on each other's alleles (Figure 7). These observations are consistent with many studies, including in allopolyploid cotton [13], *Cirsium arvense* [20], and fungal allopolyploids [54], where trans-regulatory factors play a prominent role in the regulation of hybrid expression inheritance patterns. For example, in the interspecific hybrids of *Coffea canephora* and *C. eugenioides*, allelic expression patterns, including additive and transgressive categories, have shown to depend on the combined effect of trans-regulatory factors from the parental genomes, while asymmetric effects of trans-regulatory factors appear to cause biased expression level dominance [31].

In conclusion, our data suggest that genetic divergence between hybrids and parents, cis- and trans-effects, and the gene expression patterns play important roles in establishing cabbage heterosis. The larger number of SNPs and InDels consistent with the female parent and the expression bias represented by DEGs indicate that the female parent has a higher contribution to the gene expression divergence. The expression inheritance patterns suggest a female parent expression level dominance and a mostly non-additive expression of hybrid genes. Both cis- and trans-effects tend to mediate gene expression divergence in the hybrids, but with a comparatively higher contribution from the cis-effects.

4. Materials and Methods

4.1. Plant Materials and Measurement of Horticultural Traits

Cabbage (*Brassica oleracea* L. var. *capitata*) seeds were planted in a research field at the Northwest A&F University, Shanxi, China, under normal farming conditions in late July of 2017. The female parents FP2 and FP3 belonged to the Ogura cytoplasmic male sterile (CMS) line and were obtained via seven generations of backcrossing [55]. The remaining female parent (FP1) was a new and stable cabbage variety that had been self-pollinated for six generations. The three male parents (MP1, MP2, and MP3) were double haploid lines obtained by microspore culture. These lines were selected as parents due to their high crack resistance, high yield, consistency of quality traits, and our observations that their hybrids showed superior head trait phenotypes as described in the results section. Some noticeable phenotypic differences of these lines are described in Table S2.

Plants were grown in randomized blocks with three replications per line and 20–30 plants per replication. The cabbage heads were harvested when 80% of the head leaves were at commercial maturity. In each replicate, the largest and the smallest cabbages were removed to gain a representative sample of average-sized cabbage heads. The following seven horticultural traits were evaluated: net head weight, polar head diameter, equatorial head diameter, the total weight of the distinguishable petioles (main petioles), non-wrapper leaf (photosynthetic outer leaves that are not part of the cabbage head) weight, plant height, and plant diameter (Figure S6).

MPH and HPH were calculated using the equations below [56]. Statistical significance was determined by one-way ANOVA with LSD post-hoc test (SPSS, Ver.16.0; $p < 0.05$).

$$\text{MPH (\%)} = (M_{HY} - M_{MiP})/M_{MiP} \times 100 \tag{1}$$

$$\text{HPH (\%)} = (M_{HY} - M_{HiP})/M_{HiP} \times 100 \tag{2}$$

4.2. RNA Extraction and Sequencing

The outer layer of cabbage head leaves (first layer of wrapper leaves) was used for RNA extraction when 80% of the head leaves were at commercial maturity. Approximately 10 cm^2 of the wrapper leaf was collected from the top center region, frozen immediately in liquid nitrogen, and stored at −80 °C until RNA extraction. Each sample was composed of leaves from five randomly selected cabbages. There were three biological replicates for each line. Total RNA was extracted from finely ground leaves using a TRIzol-based method (Tiangen, Beijing, China). RNA concentration and integrity were assessed on an Agilent Bioanalyzer 2100 system using the RNA Nano 6000 Assay Kit (Agilent Technologies, Chandler, USA). The RNA integrity number (RIN) was 8.1 or higher for all samples.

Sequencing libraries were generated using NEBNext Ultra RNA Library Prep Kit for Illumina (NEB, Ipswich, MA, USA), and index codes were added to match sequences to each sample. Briefly, mRNA was purified from total RNA using poly-T oligo-attached magnetic beads. Fragmentation was carried out using divalent cations under elevated temperature in NEBNext First-Strand Synthesis Reaction Buffer. The first-strand cDNA, synthesized using random hexamer primers and M-MuLV reverse transcriptase, was treated with RNase H and used for second-strand cDNA synthesis with DNA Polymerase I. The overhangs were converted into blunt ends by exonuclease. After the adenylation of 3′ ends, the DNA fragments were prepared for hybridization by ligating NEBNext adaptors bearing hairpin loop structures. The library fragments were purified with AMPure XP system (Beckman Coulter, Indianapolis, Indiana, USA) to select cDNA fragments of approximately 240 bp in length. Then, PCR was performed with Phusion High-Fidelity DNA polymerase, universal PCR primers, and Index (X) Primer. PCR products were purified (AMPure XP system), and library quality was assessed on an Agilent Bioanalyzer 2100 system. Finally, clustering of the index-coded samples was performed on a cBot Cluster Generation System, using a TruSeq PE Cluster Kit v4-cBot-HS (Illumina). Sequencing was done in an Illumina HiSeq Xten platform, and paired-end reads were generated.

4.3. Quality Control and Read Mapping

The raw data in FASTQ format were processed using in-house Perl scripts. Clean data (clean reads) were obtained by removing reads containing adapter, reads containing ploy-N, and low-quality reads from raw data. The Q20 and Q30 values, GC content, and the extent of sequence duplication level in the filtered data were determined. Thus, all downstream analyses were based on high-quality clean data. When the clean reads were mapped to the *B. oleracea* var. *capitata* reference genome (RefGen_v2.1) [57], only the reads with a perfect match or one mismatch were further analyzed and annotated. Tophat (v 2.1.1) was used to map the clean reads to the reference genome [58]. Transcriptome sequencing can be simulated as a process of random sampling, that is, to randomly extract sequence fragments from any nucleic acid sequence in a sample transcriptome. The number of fragments extracted from a gene (or transcript) obeys Beta Negative Binomial Distribution [59]. Based on this mathematical model, the Cuffquant and Cuffnorm components of the Cufflinks software were used to quantify the expression levels and genes through the location information of mapped reads on genes [58]. The expression level of each gene was calculated and normalized by the fragments per kilobase of transcript per million mapped reads (FPKM) [60].

4.4. Validation of RNA-Seq Data by qRT-PCR

To verify the accuracy of the RNA-seq data, we randomly selected ten genes from the DEGs and performed quantitative reverse transcription PCR (qRT-PCR; Table S3). The peptidyl-prolyl cis-trans isomerase (cyclophilin) gene was used as the internal reference [61]. First-strand reverse transcription was conducted using the PrimeScript RT reagent kit with gDNA Eraser (Perfect Real Time; Takara Bio., Shiga, Japan). The qRT-PCR reactions were performed with the EvaGreen 2X qPCR MasterMix Kit in an ABI 7500 Quantitative PCR System (Applied Biosystems, Foster City, CA, USA) with four technical replicates and ten samples. The cycle parameters were 95 °C for 10 min, followed by 35 amplification cycles at 60 °C for 1 min and 94 °C for 15 s. Results were analyzed using the $2^{-\Delta\Delta Ct}$ method [62]. The expression trends seen in the qPCR data were similar to that of the RNA-seq data, which confirms the reliability of the RNA-seq results (Table S4).

4.5. Analysis of the SNP Sites and the DEGs

Picard-tools (v 1.41) and SAMtools (v 0.1.18) were used to sort and remove duplicated reads, and then the bam alignment results of each sample were reordered [63]. GATK2 software was used to perform SNP and InDel analysis [64]. GATK standard filter method was used (cluster window size: 10; MQ0 $\geq$ 4; MQ0/(1.0*DP) > 0.1; QUAL < 10; QUAL < 30.0, QD < 5.0 or HRun > 5) to filter raw files, and SNPs with a set value > 5 were retained.

The DEGs among each comparison group were analyzed using the R package DESeq (v 1.10.1) [65]. The resulting P values were adjusted using the Benjamini and Hochberg's approach to control the false discovery rate. Genes with an adjusted p-value of $\leq$ 0.01 and fold change of $\geq$2 were considered as differentially expressed. GO enrichment analysis of the DEGs was implemented by the Goseq R package based on Wallenius' noncentral hypergeometric distribution. The enriched GO terms were adjusted by multiple testing (p-value < 0.05) [66].

KOBAS software was used to analyze the enrichment of DEGs in KEGG pathways [66]. Pathways with an adjusted p-value of $\leq$ 0.05 were considered as significantly enriched. The BLAST homologous sequence analysis tool and the plant transcription factor database (v4.0; http://planttfdb.cbi.pku.edu.cn/) were used to annotate the transcription factors associated with the growth and development of cabbage head leaves.

4.6. Inheritance Classification and Cis- and Trans-Regulatory Effects

Inheritance classification was performed as described in [31,33]. The DESeq package was used to normalize the expression value of parents and their hybrids. Expression inheritance was determined by subtracting the log-transformed expression values of each parent from those of the hybrids. The hybrid genes showing a total expression deviation of more than 1.25-fold from either parent were considered to have a non-conserved inheritance. Gene expressions in the hybrids that were lower than one parent but higher than the other parent were considered as showing mid-parent levels and were classified as additive. Genes showing expression levels similar to one parent were classified as dominant (maternal or paternal and high-parent or low-parent dominance). Hybrid gene expressions greater or less than both parents were classified as showing transgressive-up regulations and transgressive-down regulations, respectively [31,33].

To infer hybrid ASE levels, parent-specific SNPs were identified using custom Perl scripts. Only the divergent polymorphic nucleotide sites where accessions of both parents were homozygous for a given difference were retained. SNPs with a minimum of 10x read coverage in the hybrids were used to determine allele-specific expressions (ASEs) and to distinguish paternal and maternal alleles in the hybrids. For the quantification of ASE, the DESeq package was used to normalize mapped read depth coverage at SNP sites in the hybrid and parental alignments. After applying quality control criteria, 31,962, 52,177, 34,653, and 56,099 SNPs and 7223, 9246, 6937, and 9628 genes with ≥ 10 read coverage were identified in HY1–HY4, respectively. Relative ASE was calculated as the percentage of allele-specific read counts representing the female parent (Fisher's exact test, p-value < 0.05) [31].

The cis- and trans-regulatory effects on the gene expression of hybrids were determined using ASE. Since the parental alleles in a hybrid are in the same cellular environment and share the same trans-regulatory factors, there is no trans-regulatory effect on the expression differences between parental alleles; therefore, the ASE divergence in the hybrids reflects the cis-effect. The trans-effect could be estimated by subtracting the cis-effect from the overall expression divergence between male and female parents [21,32]. The cis-effects were determined by evaluating the ASE ratios in a given hybrid (two-sided prop. test in R, H_0: $FP_{HY}/MP_{HY} = 1$, Benjamini–Hochberg method). The trans-effects were determined by comparing parental expression ratio with the allelic expression ratio of a given hybrid (H_0: $FP/MP = FP_{HY}/MP_{HY}$, Benjamini–Hochberg method).

Supplementary Materials: The following are available online at https://www.mdpi.com/2223-7747/10/2/275/s1. Figure S1. GO pathway enrichment analysis for DEGs of FP2 vs. HY3 and MP2 vs. HY3. The dark color bars show percentages/numbers of DEGs representing each category. The light color bars show the percentage/numbers of the GO categories of all the expressed genes detected. Figure S2. COG and KOG pathway enrichment analysis of DEGs for FP2 vs. HY3 and MP2 vs. HY3. (A,B) COG pathway enrichment analysis of DEGs for FP2 vs. HY3 and MP2 vs. HY3, respectively. (C,D) KOG pathway enrichment analysis of DEGs for FP2 vs. HY3 and MP2 vs. HY3, respectively. Figures S3–S5. Relative allele-specific expressions (ASEs) and cis- and trans-effects of HY2, HY3, and HY4, respectively. The relative ASE represents the expression of maternal alleles as a percentage of total gene expression in the hybrid (%FPHY). The pie chart shows the proportion of genes showing cis-effects, trans-effects, or both cis- and trans- (cis + trans) effects. Genes showing no significant evidence of cis- or trans- effects were classified as conserved. Figure S6. Horticultural traits of the cabbage head. The non-wrapper leaves represent the photosynthetic outer leaves that are not part of the cabbage head. The distinguishable petioles of each layer were cut out to assess the total weight of the main petioles (shown in white arrows). Table S1. Selected transcription factors represented by DEGs that are likely involved in the growth and development of cabbage heads. Table S2. Backgrounds of the six parental inbred lines. Table S3. List of the forward (F) and reverse (R) primer combinations used in the qRT-PCR analysis. Table S4. Comparison of qRT-PCR and RNA-Seq expression data.

Author Contributions: Conceptualization, Z.X. and E.Z.; methodology, J.G.; software, J.G.; validation, S.L.; formal analysis, S.L. and C.P.A.J.; investigation, S.L.; resources, Z.X.; data curation, X.W.; writing—original draft preparation, S.L.; writing—review and editing, C.P.A.J.; visualization, S.L. and C.P.A.J.; supervision, E.Z.; project administration, X.W.; funding acquisition, Z.X. All authors have read and agreed to the published version of the manuscript.

Funding: This research was funded by National Key Research and Development Program of China (2016YFD0101702, 2017YFD 0101804), National Technical System of Bulk Vegetable Industry (CARS-23-G22), and Key Research and Development Programs in Shaanxi (2018NY-059).

Data Availability Statement: The sequencing reads generated in this study were deposited in the NCBI Sequence Read Archive (SRA) under accession number PRJNA664256.

Acknowledgments: We would like to thank Leonardo Galindo-González, University of Alberta, Canada, for his valuable comments.

Conflicts of Interest: The authors declare no conflict of interest.

References

1. Ryder, P.; McKeown, P.C.; Fort, A.; Spillane, C. Epigenetics and heterosis in crop plants. In *Epigenetics in Plants of Agronomic Importance: Fundamentals and Applications*; Springer: Cham, Switzeerland, 2014; pp. 129–147. [CrossRef]
2. Gallais, A. Heterosis: Its genetic basis and its utilisation in plant breeding. *Euphytica* **1988**, *39*, 95–104. [CrossRef]
3. Rokayya, S.; Li, C.-J.; Zhao, Y.; Li, Y.; Sun, C.-H. Cabbage (*Brassica oleracea* L. var. *capitata*) phytochemicals with antioxidant and anti-inflammatory potential. *Asian Pac. J. Cancer Prev.* **2013**, *14*, 6657–6662. [CrossRef] [PubMed]
4. Hochholdinger, F.; Hoecker, N. Towards the molecular basis of heterosis. *Trends Plant Sci.* **2007**, *12*, 427–432. [CrossRef] [PubMed]
5. Chen, Z.J. Genomic and epigenetic insights into the molecular bases of heterosis. *Nat. Rev. Genet.* **2013**, *14*, 471–482. [CrossRef] [PubMed]
6. Goff, S.A. A unifying theory for general multigenic heterosis: Energy efficiency, protein metabolism, and implications for molecular breeding. *New Phytol.* **2011**, *189*, 923–937. [CrossRef] [PubMed]
7. Fiévet, J.B.; Nidelet, T.; Dillmann, C.; De Vienne, D. Heterosis is a systemic property emerging from non-linear genotype-phenotype relationships: Evidence from in vitro genetics and computer simulations. *Front. Genet.* **2018**, *9*, 159. [CrossRef] [PubMed]
8. Kaeppler, S. Heterosis: One boat at a time, or a rising tide? *New Phytol.* **2011**, *189*, 900–902. [CrossRef]
9. Fu, D.; Xiao, M.; Hayward, A.; Jiang, G.; Zhu, L.; Zhou, Q.; Li, J.; Zhang, M. What is crop heterosis: New insights into an old topic. *J. Appl. Genet.* **2015**, *56*, 1–13. [CrossRef]
10. Li, D.; Zeng, R.; Li, Y.; Zhao, M.; Chao, J.; Li, Y.; Wang, K.; Zhu, L.; Tian, W.-M.; Liang, C. Gene expression analysis and SNP/InDel discovery to investigate yield heterosis of two rubber tree F1 hybrids. *Sci. Rep.* **2016**, *6*, 1–12. [CrossRef]
11. Hu, X.; Wang, H.; Diao, X.; Liu, Z.; Li, K.; Wu, Y.; Liang, Q.; Wang, H.; Huang, C. Transcriptome profiling and comparison of maize ear heterosis during the spikelet and floret differentiation stages. *BMC Genom.* **2016**, *17*, 959. [CrossRef]
12. Swanson-Wagner, R.A.; Jia, Y.; DeCook, R.; Borsuk, L.A.; Nettleton, D.; Schnable, P.S. All possible modes of gene action are observed in a global comparison of gene expression in a maize F1 hybrid and its inbred parents. *Proc. Natl. Acad. Sci. USA* **2006**, *103*, 6805–6810. [CrossRef] [PubMed]
13. Yoo, M.-J.; Szadkowski, E.; Wendel, J.F. Homoeolog expression bias and expression level dominance in allopolyploid cotton. *Heredity* **2013**, *110*, 171–180. [CrossRef] [PubMed]
14. Dapp, M.; Reinders, J.; Bédiée, A.; Balsera, C.; Bucher, E.; Theiler, G.; Granier, C.; Paszkowski, J. Heterosis and inbreeding depression of epigenetic Arabidopsis hybrids. *Nat. Plants* **2015**, *1*, 15092. [CrossRef] [PubMed]
15. Tian, M.; Nie, Q.; Li, Z.; Zhang, J.; Liu, Y.; Long, Y.; Wang, Z.; Wang, G.; Liu, R. Transcriptomic analysis reveals overdominance playing a critical role in nicotine heterosis in *Nicotiana tabacum* L. *BMC Plant Biol.* **2018**, *18*, 1–10. [CrossRef]
16. Guo, M.; Rupe, M.A.; Zinselmeier, C.; Habben, J.; Bowen, B.A.; Smith, O.S. Allelic variation of gene expression in maize hybrids. *Plant Cell.* **2004**, *16*, 1707–1716. [CrossRef]
17. Springer, N.M.; Stupar, R.M. Allele-specific expression patterns reveal biases and embryo-specific parent-of-origin effects in hybrid maize. *Plant Cell.* **2007**, *19*, 2391–2402. [CrossRef]
18. Fontanillas, P.; Landry, C.R.; Wittkopp, P.J.; Russ, C.; Gruber, J.D.; Nusbaum, C.; Hartl, D.L. Key considerations for measuring allelic expression on a genomic scale using high-throughput sequencing. *Mol. Ecol.* **2010**, *19*, 212–227. [CrossRef]
19. Stupar, R.M.; Springer, N.M. Cis-transcriptional variation in maize inbred lines B73 and Mo17 leads to additive expression patterns in the F1 hybrid. *Genetics* **2006**, *173*, 2199–2210. [CrossRef]
20. Bell, G.D.; Kane, N.C.; Rieseberg, L.H.; Adams, K.L. RNA-Seq analysis of allele-specific expression, hybrid effects, and regulatory divergence in hybrids compared with their parents from natural populations. *Genome Biol. Evol.* **2013**, *5*, 1309–1323. [CrossRef]
21. Shi, X.; Ng, D.W.-K.; Zhang, C.; Comai, L.; Ye, W.; Chen, Z.J. Cis- and trans-regulatory divergence between progenitor species determines gene-expression novelty in *Arabidopsis* allopolyploids. *Nat. Commun.* **2012**, *3*, 950–959. [CrossRef]
22. Singh, B.; Sharma, S.; Singh, B. Heterosis for mineral elements in single cross-hybrids of cabbage (*Brassica oleracea* var. *capitata* L.). *Sci. Hortic.* **2009**, *122*, 32–36. [CrossRef]
23. Singh, B.K.; Sharma, S.R.; Singh, B. Genetic combining ability for concentration of mineral elements in cabbage head (*Brassica oleracea* var. *capitata* L.). *Euphytica* **2012**, *184*, 265–273. [CrossRef]
24. Zhang, X.; Su, Y.; Liu, Y.; Fang, Z.; Yang, L.; Zhuang, M.; Zhang, Y.; Li, Z.; Lv, H. Genetic analysis and QTL mapping of traits related to head shape in cabbage (*Brassica oleracea* var. *capitata* L.). *Sci. Hortic.* **2016**, *207*, 82–88. [CrossRef]
25. Broadley, M.R.; Hammond, J.P.; King, G.J.; Astley, D.; Bowen, H.C.; Meacham, M.C.; Mead, A.; Pink, D.A.; Teakle, G.R.; Hayden, R.M.; et al. Shoot calcium and magnesium concentrations differ between subtaxa, are highly heritable, and associate with potentially pleiotropic loci in *Brassica oleracea*. *Plant Physiol.* **2008**, *146*, 1707–1720. [CrossRef] [PubMed]
26. Jeong, S.-Y.; Ahmed, N.U.; Jung, H.-J.; Kim, H.-T.; Park, J.-I.; Nou, I.S. Discovery of candidate genes for heterosis breeding in *Brassica oleracea* L. *Acta Physiol. Plant* **2017**, *39*, 1–12. [CrossRef]
27. Moon, J.; Hake, S. How a leaf gets its shape. *Curr. Opin. Plant Biol.* **2011**, *14*, 24–30. [CrossRef]
28. Dubos, C.; Stracke, R.; Grotewold, E.; Weisshaar, B.; Martin, C.; Lepiniec, L. MYB transcription factors in *Arabidopsis*. *Trends Plant Sci.* **2010**, *15*, 573–581. [CrossRef]
29. Blein, T.; Hasson, A.; Laufs, P. Leaf development: What it needs to be complex. *Curr. Opin. Plant Biol.* **2010**, *13*, 75–82. [CrossRef]
30. Krogan, N.T.; Berleth, T. The identification and characterization of specific ARF-Aux/IAA regulatory modules in plant growth and development. *Plant Signal Behav.* **2015**, *10*, e992748. [CrossRef]

31. Combes, M.-C.; Hueber, Y.; Dereeper, A.; Rialle, S.; Herrera, J.-C.; Lashermes, P. Regulatory divergence between parental alleles determines gene expression patterns in hybrids. *Genome Biol. Evol.* **2015**, *7*, 1110–1121. [CrossRef]
32. Wittkopp, P.J.; Haerum, B.K.; Clark, A.G. Evolutionary changes in cis and trans gene regulation. *Nature* **2004**, *430*, 85–88. [CrossRef] [PubMed]
33. Mcmanus, C.J.; Coolon, J.D.; Duff, M.O.; Eipper-Mains, J.; Graveley, B.R.; Wittkopp, P.J. Regulatory divergence in *Drosophila* revealed by mRNA-seq. *Genome Res.* **2010**, *20*, 816–825. [CrossRef] [PubMed]
34. Wei, G.; Tao, Y.; Liu, G.; Chen, C.; Luo, R.; Xia, H.; Gan, Q.; Zeng, H.; Lu, Z.; Han, Y.; et al. A transcriptomic analysis of superhybrid rice LYP9 and its parents. *Proc. Natl. Acad. Sci. USA* **2009**, *106*, 7695–7701. [CrossRef] [PubMed]
35. Huang, X.; Yang, S.; Gong, J.; Zhao, Q.; Feng, Q.; Zhan, Q.; Zhao, Y.; Li, W.; Cheng, B.; Xia, J.; et al. Genomic architecture of heterosis for yield traits in rice. *Nature* **2016**, *537*, 629–633. [CrossRef]
36. Botet, R.; Keurentjes, J.J. The role of transcriptional regulation in hybrid vigor. *Front. Plant Sci.* **2020**, *11*, 410. [CrossRef]
37. Gehring, M. Genomic imprinting: Insights from plants. *Annu. Rev. Genet.* **2013**, *47*, 187–208. [CrossRef]
38. Guo, M.; Rupe, M.A.; Yang, X.; Crasta, O.; Zinselmeier, C.; Smith, O.S.; Bowen, B. Genome-wide transcript analysis of maize hybrids: Allelic additive gene expression and yield heterosis. *Theor. Appl. Genet.* **2006**, *113*, 831–845. [CrossRef]
39. Lisec, J.; Steinfath, M.; Meyer, R.C.; Selbig, J.; Melchinger, A.E.; Willmitzer, L.; Altmann, T. Identification of heterotic metabolite QTL in *Arabidopsis thaliana* RIL and IL populations. *Plant J.* **2009**, *59*, 777–788. [CrossRef]
40. Zhang, T.-F.; Li, B.; Zhang, D.-F.; Jia, G.-Q.; Li, Z.-Y.; Wang, S. Genome-wide transcriptional analysis of yield and heterosis-associated genes in maize (*Zea mays* L.). *J. Integr. Agric.* **2012**, *11*, 1245–1256. [CrossRef]
41. Zhai, R.; Feng, Y.; Wang, H.; Zhan, X.; Shen, X.; Wu, W.; Cao, L.-Y.; Chen, D.; Dai, G.; Yang, Z.; et al. Transcriptome analysis of rice root heterosis by RNA-Seq. *BMC Genom.* **2013**, *14*, 19. [CrossRef]
42. Zhang, H.-Y.; He, H.; Chen, L.; Li, L.; Liang, M.-Z.; Wang, X.; Liu, X.-G.; He, G.-M.; Chen, R.-S.; Ma, L.; et al. A genome-wide transcription analysis reveals a close correlation of promoter INDEL polymorphism and heterotic gene expression in rice hybrids. *Mol. Plant* **2008**, *1*, 720–731. [CrossRef] [PubMed]
43. Li, B.; Zhang, D.-F.; Jia, G.-Q.; Dai, J.-R.; Wang, S.-C. Genome-wide comparisons of gene expression for yield heterosis in maize. *Plant Mol. Biol. Rep.* **2009**, *27*, 162–176. [CrossRef]
44. Groszmann, M.; Gonzalez-Bayon, R.; Lyons, R.L.; Greaves, I.K.; Kazan, K.; Peacock, W.J.; Dennis, E.S. Hormone-regulated defense and stress response networks contribute to heterosis in *Arabidopsis* F1 hybrids. *Proc. Natl. Acad. Sci. USA* **2015**, *112*, E6397–E6406. [CrossRef] [PubMed]
45. Rédei, G.P. Single locus heterosis. *Mol. Gen. Genet.* **1962**, *93*, 164–170.
46. Ma, J.; Li, J.; Cao, Y.; Wang, L.; Wang, F.; Wang, H.; Li, H.-Y. Comparative study on the transcriptome of maize mature embryos from two China elite hybrids Zhengdan958 and Anyu5. *PLoS ONE* **2016**, *11*, e0158028. [CrossRef]
47. Taliercio, E.; Eickholt, D.; Rouf, R.; Carter, T. Changes in gene expression between a soybean F1 hybrid and its parents are associated with agronomically valuable traits. *PLoS ONE* **2017**, *12*, e0177225. [CrossRef]
48. Zhang, J.; Li, G.; Li, H.; Pu, X.; Jiang, J.; Chai, L.; Zheng, B.; Cui, C.; Yang, Z.; Zhu, Y.; et al. Transcriptome analysis of interspecific hybrid between *Brassica napus* and *B. rapa* reveals heterosis for oil rape improvement. *Int. J. Genom.* **2015**, *2015*, 230985.
49. Jindal, S.K.; Thakur, J.C. Genetic architecture of quantitative characters in November maturity cauliflower. *Crop Improv.* **2005**, mboxemph32, 92–94.
50. Meyer, R.C.; Witucka-Wall, H.; Becher, M.; Blacha, A.; Boudichevskaia, A.; Dörmann, P.; Fiehn, O.; Friedel, S.; Von Korff, M.; Lisec, J.; et al. Heterosis manifestation during early *Arabidopsis* seedling development is characterized by intermediate gene expression and enhanced metabolic activity in the hybrids. *Plant J.* **2012**, *71*, 669–683. [CrossRef]
51. Lemos, B.; Araripe, L.O.; Fontanillas, P.; Hartl, D.L. Dominance and the evolutionary accumulation of cis- and trans-effects on gene expression. *Proc. Natl. Acad. Sci. USA* **2008**, *105*, 14471–14476. [CrossRef]
52. Vuylsteke, M.; Van Eeuwijk, F.; Van Hummelen, P.; Kuiper, M.; Zabeau, M. Genetic analysis of variation in gene expression in *Arabidopsis thaliana*. *Genetics* **2005**, *171*, 1267–1275. [CrossRef] [PubMed]
53. A Signor, S.; Nuzhdin, S.V. The evolution of gene expression in cis and trans. *Trends Genet.* **2018**, *34*, 532–544. [CrossRef] [PubMed]
54. Cox, M.P.; Dong, T.; Shen, G.; Dalvi, Y.B.; Scott, D.B.; Ganley, A.R.D. An interspecific fungal hybrid reveals cross-kingdom rules for allopolyploid gene expression patterns. *PLoS Genet.* **2014**, *10*, e1004180. [CrossRef] [PubMed]
55. Hiroshi, O. Studies on the new male-sterility in Japanese radish, with special reference to the utilization of this sterility towards the practical raising of hybrid seeds. *Mem. Fac. Agric. Kagoshima Univ.* **1968**, *6*, 40–75.
56. Peng, Y.; Shi, D.; Zhang, T.; Li, X.; Fu, T.; Xu, Y.; Wan, Z. Development and utilization of an efficient cytoplasmic male sterile system for Cai-xin (*Brassica rapa* L.). *Sci. Hortic.* **2015**, *190*, 36–42. [CrossRef]
57. Liu, S.; Liu, Y.; Yang, X.; Tong, C.; Edwards, D.; Parkin, I.A.P.; Zhao, M.; Ma, J.; Yu, J.; Huang, S.; et al. The *Brassica oleracea* genome reveals the asymmetrical evolution of polyploid genomes. *Nat. Commun.* **2014**, *5*, 1–11. [CrossRef] [PubMed]
58. Ghosh, S.; Chan, C.-K.K. *Analysis of RNA-Seq Data Using TopHat and Cufflinks*; Humana Press Inc.: Totowa, NJ, USA, 2016; Volume 1374, pp. 339–361.
59. Jiang, H.; Wong, W. Statistical inferences for isoform expression in RNA-Seq. *Bioinformatics* **2009**, *25*, 1026–1032. [CrossRef]
60. Sims, D.W.; Sudbery, I.M.; Ilott, N.E.; Heger, A.; Ponting, C.P. Sequencing depth and coverage: Key considerations in genomic analyses. *Nat. Rev. Genet.* **2014**, *15*, 121–132. [CrossRef]

61. Kumar, V.; Sharma, R.; Trivedi, P.C.; Vyas, G.K.; Khandelwal, V. Traditional and novel references towards systematic normalization of qRT-PCR data in plants. *Aust. J. Crop Sci.* **2011**, *5*, 1455–1468.
62. Adnan, M.; Morton, G.; Hadi, S. Analysis of *rpoS* and *bolA* gene expression under various stress-induced environments in planktonic and biofilm phase using $2^{-\Delta\Delta CT}$ method. *Mol. Cell Biochem.* **2011**, *357*, 275–282. [CrossRef]
63. Li, H.; Handsaker, B.; Wysoker, A.; Fennell, T.; Ruan, J.; Homer, N.; Marth, G.; Abecasis, G.; Durbin, R. The sequence alignment/map (SAM) format and SAMtools. *Bioinformatics* **2009**, *25*, 2078–2079. [CrossRef] [PubMed]
64. McKenna, A.; Hanna, M.; Banks, E.; Sivachenko, A.; Cibulskis, K.; Kernytsky, A.; Garimella, K.; Altshuler, D.; Gabriel, S.B.; Daly, M.J.; et al. The genome analysis toolkit: A mapreduce framework for analyzing next-generation DNA sequencing data. *Genome Res.* **2010**, *20*, 1297–1303. [CrossRef] [PubMed]
65. Anders, S.; Huber, W. Differential expression analysis for sequence count data. *Genome Biol.* **2010**, *11*, 1. [CrossRef] [PubMed]
66. Mao, X.; Cai, T.; Olyarchuk, J.G.; Wei, L. Automated genome annotation and pathway identification using the KEGG Orthology (KO) as a controlled vocabulary. *Bioinformatics* **2005**, *21*, 3787–3793. [CrossRef]

Article

Genome-Wide Identification and Analysis of *SRO* Gene Family in Chinese Cabbage (*Brassica rapa* L.)

Yali Qiao [1], Xueqin Gao [1], Zeci Liu [1], Yue Wu [1], Linli Hu [1,2,*] and Jihua Yu [1,2,*]

[1] College of Horticulture, Gansu Agricultural University, Lanzhou 730070, China; qiaoyl@st.gsau.edu.cn (Y.Q.); gaoxq@st.gsau.edu.cn (X.G.); liuzc@gsau.edu.cn (Z.L.); wuy@gsau.edu.cn (Y.W.)

[2] Gansu Provincial Key Laboratory of Aridland Crop Science, Gansu Agricultural University, Lanzhou 730070, China

* Correspondence: hull@gsau.edu.cn (L.H.); yujihua@gsau.edu.cn (J.Y.); Tel.: +86-931-7632188 (J.Y.)

Received: 6 August 2020; Accepted: 16 September 2020; Published: 18 September 2020

Abstract: Similar to radical-induced cell death 1 (SROs) is a family of small proteins unique to plants. *SRO* transcription factors play an important role in plants' response to biotic and abiotic stresses. In this study, we identified 12 *BrSRO* genes in Chinese cabbage (*Brassica rapa* L.). Among them, a comprehensive overview of the *SRO* gene family is presented, including physical and chemical characteristics, chromosome locations, phylogenetic analysis, gene structures, motif analysis, and cis-element analyses. The number of amino acids of *BrSRO* genes is between 77–779 aa, isoelectric point changed from 6.02 to 9.6. Of the 12 *BrSRO* genes, 11 were randomly distributed along the 7 chromosomes, while *BrSRO12* was located along unassigned scaffolds. Phylogenetic analysis indicated that the SRO proteins from six species, including *Arabidopsis*, banana, rice, *Solanum lycopersicum*, *Zea mays*, and Chinese cabbage were divided into eleven groups. The exon-rich *BrSRO6* and *BrSRO12* containing 15 exons were clustered to group K. All 12 genes have motif 2, which indicate that motif 2 is a relatively conservative motif. There are many hormone and stress response elements in *BrSRO* genes. The relative expression levels of 12 *BrSRO* genes under high temperature, drought, salt, and low temperature conditions were analyzed by real-time fluorescence quantitative PCR. The results indicated the relative expression level of *BrSRO8* was significantly up-regulated when plants were exposed to high temperature. The relative expression levels of *BrSRO1, 3, 7, 8,* and *9* were higher under low temperature treatment. The up-regulated genes response to drought and salt stresses were *BrSRO1, 5, 9* and *BrSRO1, 8,* respectively. These results indicated that these genes have certain responses to different abiotic stresses. This work has provided a foundation for further functional analyses of *SRO* genes in Chinese cabbage.

Keywords: Chinese cabbage; *SRO* gene family; abiotic stress; bioinformatics; expression analysis

1. Introduction

When subjected to stresses, plants can survive in complex and diverse environments for stress-induced gene expression. In these processes, similar to radical-induced cell death 1 proteins (SROs) participate in multiple regulatory networks through stress response [1,2]. *SRO* is a family of small proteins unique to plants. It plays an important role in plant growth and development and in responding to abiotic stresses, such as salt, drought, and heavy metals. *SROs* generally contain a poly(ADP-ribose) polymerase catalytic (PARP, PS51059) center and a RCD1-SRO-TAF4(RST, PF12174) conservative domain [3], part of the *SROs* also contains N-terminal WWE (PS50918) domain [4]. In *Arabidopsis*, there are six members in the *AtSRO* family, namely *AtRCD1* and *AtSRO1-5* [5]. *AtRCD1* is the first member of the *SRO* family identified in *Arabidopsis* [6]. *AtRCD1* can interact with transcription factors in the nucleus to participate in the drought response mediated by the plant abscisic acid signaling pathway, and can also regulate plant development through hormone signaling pathways including abscisic acid (ABA), ethylene (ETH), methyl jasmonate (MEJA) and so on [7,8]. *AtSRO1* and *AtRCD1*, two homologous genes, have functional redundancy under different stress conditions [9]. *AtSRO1* is involved in abiotic stress response, and its mutant *SRO1-1* has strong resistance to osmotic and oxidative stress [3,10]. *AtSRO5* interacts with transcription factors to regulate gene expression, and overexpression of *AtSRO5* can increase the salt tolerance of transgenic plants by lowering the level of H_2O_2 in the roots [11]. *AtSRO2* and *AtSRO3* can respond to strong light, salt, and ozone stress; *AtSRO4* has no clear function reported [12]. *SROs* have also been partially studied in apple, rice, wheat, corn, continental cotton, tomato, and other crops. For example, in apple, *MdRCD1* can regulate the pore size through ABA signaling pathway, tolerate drought stress, and regulate root growth [13]. In rice, *OsSRO1c* participates in drought and oxidative stress through promoting stomatal closure and H_2O_2 accumulation by regulating SNAC1 and DST [14]. In wheat, *Ta-SRO1* can improve drought tolerance by regulating REDOX balance in plants [1]. The *SRO* gene families in various species have been identified, and the mechanism of *SROs* in response to drought stress is becoming increasingly clear.

Although a large number of studies on *SRO* genes in various species have been conducted, studies on *SRO* genes of Chinese cabbage have still not been reported. Chinese cabbage (*Brassica rapa* L.), which originated from China, is one of the specialty vegetables in the country. Chinese cabbage is rich in a variety of nutrients and is loved by consumers. Leaf bulb is the main edible part of Chinese cabbage. The growth and development of each organ of Chinese cabbage directly affect the development of leaf bulb, and then affect the yield and quality of Chinese cabbage. The development of Chinese cabbage is controlled by both gene and environment. The completion of genome sequencing of Chinese cabbage in 2011 [15] provided important reference information for bioinformatics analysis, genetic breeding, and key functional gene mining of Chinese cabbage gene family system at the whole genome level.

At present, multiple gene families of Chinese cabbage such as *HSF* [16], *AQP* [17], *TCP* [18], and *MYB* [19] have been identified by bioinformatics methods, and some genes have also been functionally verified. However, the identification and expression pattern response to various stresses of *SRO* gene families in Chinese cabbage have not been reported until now. Therefore, in this study, based on whole genome sequencing results, the members of *SRO* gene family in Chinese cabbage were identified via a bioinformatics analysis method, and subsequently the physical and chemical properties, evolutionary characteristics of its members, and protein structure were analyzed. Finally, the expression pattern of *BrSROs*' response to high temperature, low temperature, drought, and salt stress were set up via real-time quantification PCR methods. Our study provides a foundation for further research on the molecular mechanism of *SRO* gene mediating physiological growth process and stress response, and a significant basis for the genetic improvement of Chinese cabbage.

2. Result

2.1. Identification and Chromosomal Location of the SRO Family Genes in Chinese Cabbage

In this study, a total of 12 *BrSRO* genes were identified in the genome network of Chinese cabbage (Table 1). All genes were named respectively from *BrSRO1* to *BrSRO12* according to their position from the top to the bottom of Chinese cabbage chromosomes A02–A09. The number of amino acids of *BrSRO* genes is between 77–779 aa, with *BrSRO12* encoding the longest protein and highest molecular weight (85,523.47) and *BrSRO1* encoding the shortest protein and lowest molecular weight (8830.55). Furthermore, the isoelectric point changed from 6.02 (*BrSRO10*) to 9.6 (*BrSRO1*) and instability index changed from 33.9 (*BrSRO1*) to 59.1 (*BrSRO2*). *BrSRO1* has the largest fat index (100.13); the fat indexes of the rest *BrSRO* genes are between 61.69 and 88.16. In addition, the protein subcellular localization prediction showed that *BrSRO1* and *BraSRO9* proteins were predictably located in the chloroplast and nucleus. *BrSRO2*, *BrSRO4*, *BrSRO7*, and *BrSRO11* were predictably located in the chloroplast. The remaining genes were predicted to be located in the nucleus.

Table 1. Physiochemical characteristics of identified BrSRO genes.

Gene Name	Gene ID	Chromosome Location	Protein Length (aa)	Molecular Weight (kd)	PI	Total Number of Atoms	INSTABILITY Index	Fat Index	Predicted Subcellular Localization
BrSRO1	Bra033139	Chromosome A02: 16,846,412–16,846,733	77	8830.55	9.6	1272	33.9	100.13	Chloroplast. Nucleus.
BrSRO2	Bra029254	Chromosome A02: 26,296,663–26,297,969	303	34,150.08	8.68	4793	59.1	82.71	Chloroplast.
BrSRO3	Bra017317	Chromosome A04: 15,393,395–15,395,430	530	58,637.64	6.99	8192	34.19	80.55	Nucleus.
BrSRO4	Bra005336	Chromosome A05: 4,905,877–4,908,322	524	58,577.61	6.1	8178	37.71	80.29	Chloroplast.
BrSRO5	Bra010096	Chromosome A06: 19,383,958–19,385,124	313	34,876.77	8.59	4889	57.49	80	Nucleus.
BrSRO6	Bra033662	Chromosome A06: 25,753,358–25,757,833	771	85,056.17	9.07	11,856	57.48	63.97	Nucleus.
BrSRO7	Bra012380	Chromosome A07: 8,098,261–8,099,752	310	34,575.22	8.86	4865	45.5	85.29	Chloroplast.
BrSRO8	Bra016219	Chromosome A07: 18,821,147–18,822,313	304	33,789.47	8.15	4744	39	88.16	Chloroplast.
BrSRO9	Bra035511	Chromosome A08: 7,983,100–7,985,345	558	62,697.49	6.59	8754	37.36	79	Chloroplast. Nucleus.
BrSRO10	Bra023252	Chromosome A09: 20,223,502–20,225,770	482	54,230.68	6.02	7557	42.34	78.28	Nucleus.
BrSRO11	Bra024609	Chromosome A09: 24,077,869–24,079,029	308	33,418.73	6.19	4682	40.86	84.84	Chloroplast.
BrSRO12	Bra035961	Scaffold000111: 11,933–15,826	779	85,523.47	8.98	11,902	55.21	61.69	Nucleus.

The identified 12 *SRO* genes in Chinese cabbage were mapped onto chromosomes or scaffolds. Among these, 11 genes (*BrSRO1-11*) were located in chromosomes, whereas the *BrSRO12* were distributed in unmapped scaffolds (Figure 1). In detail, the 11 predicted *BrSROs* were distributed unevenly across its 7 chromosomes. Each of chromosomes A02, 06, 07, and 09 harbored two *BrSRO* genes, and a single *BrSRO* gene was located in each of the chromosomes A04, 05, and 08.

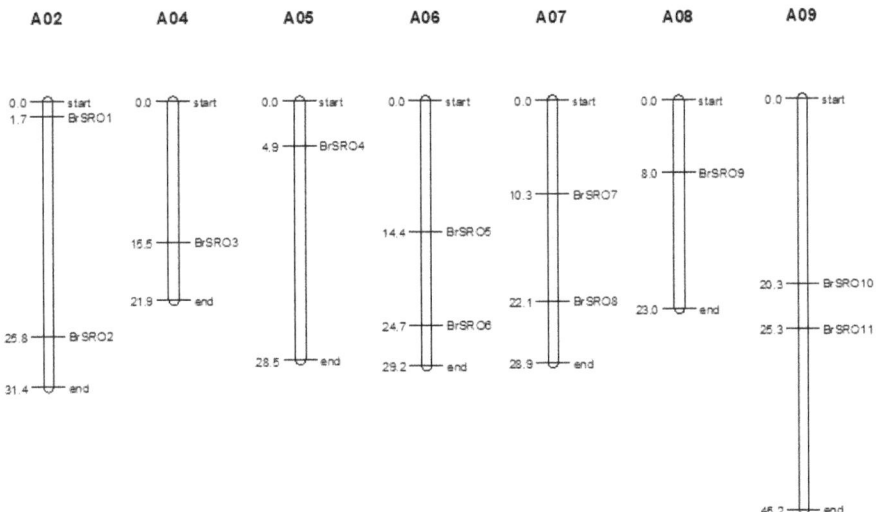

Figure 1. The chromosomal mapping analysis of the *SRO* gene family in Chinese cabbage. The chromosome number (A02–A09) is indicated at the top of each chromosome. The numbers on the left of each chromosome represent the initial position of each gene.

2.2. Phylogenetic Analysis of the SRO Family Genes in Chinese Cabbage

The SRO proteins in Chinese cabbage were compared with other species to investigate the evolutionary relationships of SRO proteins. A phylogenetic tree was constructed on the basis of 40 putative nonredundant SRO protein sequences from six species, including *Arabidopsis*, banana, rice, *Solanum lycopersicum*, *Zea mays* and Chinese cabbage (Figure 2). All 40 SRO proteins were clustered into eleven groups (A–K), which consisted 6, 2, 1, 5, 4, 2, 2, 5, 5, 5, and 2 members, respectively. All *BrSROs* were clustered into Group A, H, I, and K, which indicated that the *SRO* of Chinese cabbage gene has higher homology with the *Arabidopsis* and tomato genes, compared with rice, maize and banana. The low bootstrap values in the tree are due to divergent SRO protein sequence among *Arabidopsis*, banana, rice, *Solanum lycopersicum*, *Zea mays*, and Chinese cabbage. This is not surprising, given that both *A. thaliana* and *B. rapa* belong to cruciferous plants, and the *SRO* genes in these two species were clustered together.

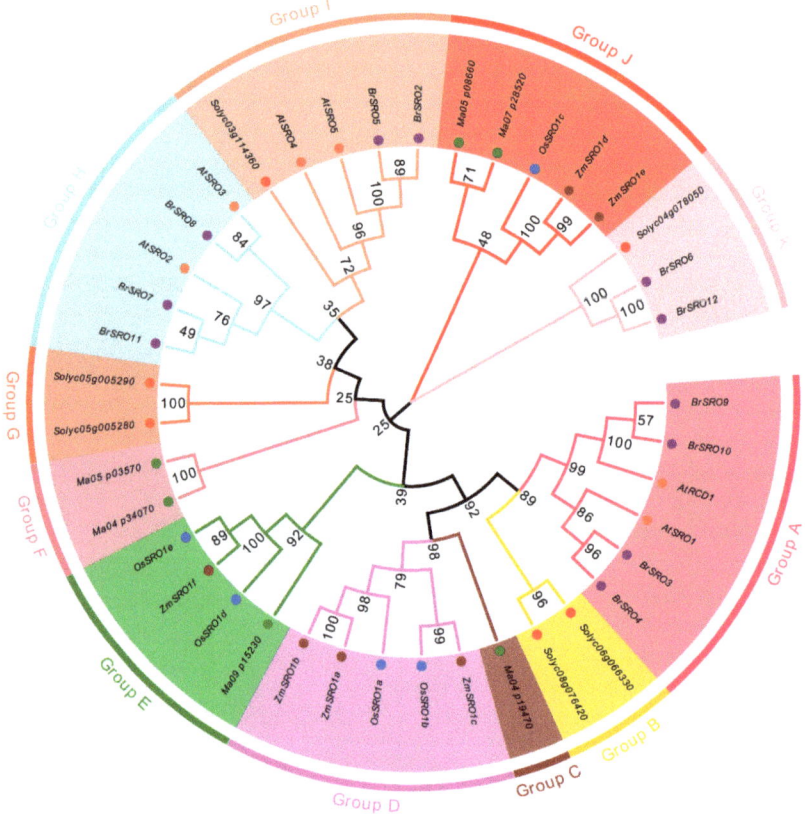

Figure 2. Phylogenetic tree of the similar to radical-induced cell death 1 (*SRO*) genes from *Arabidopsis thaliana* (At), *Solanum Lycopersicum* (Sl), *Brassica rapa* (Br), *O. sativa* (Os), *Zea mays* (Zm), *Musa acuminate* (Ma). In total, 6 AtSROs, 6 SlSROs, 11 BrSROs, 5 OsSROs, 6 ZmSROs, and 5 MaSROs were included. The phylogenetic tree was constructed for the SRO protein sequences in *Arabidopsis thaliana* (tomato), *Solanum Lycopersicum* (red), *Brassica rapa* (purple), *Oryza sativa* (blue), *Zea mays* (darkred) and *Musa acuminate* (green) using the Maximum-Likelihood method in MEGA 7.0. Bootstrap values from 1000 replicates are displayed at each node. The proteins on the tree can be divided into 11 groups from Group A to Group K, and the different groups are indicated by different colors.

2.3. Gene Structure of the BrSRO Genes

The predicted exon–intron structure was analyzed to gain an insight into the variation of the *SRO* genes in Chinese cabbage. On the basis of the evolutionary relationships of the Chinese cabbage phylogenetic tree (Figure 3a), the structure features were determined (Figure 3b). Phylogenetic analysis indicated that 12 *BrSRO* family members were divided into four groups (A, H, I, and K). All of the 12 *BrSRO* genes have complete gene structure. Interestingly, the exon-rich *BrSRO* genes containing 15 exons were clustered in group K, while the number of exons in the rest of groups ranged from 2 to 5, and the exon number of *BrSRO1* was the lowest.

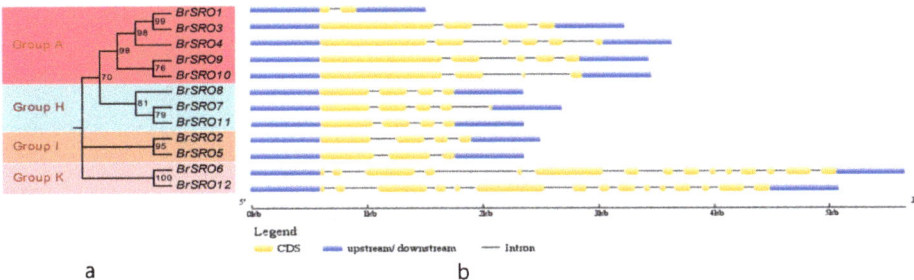

Figure 3. Compositions of introns and exons of *BrSRO* genes based on the phylogenetic relations. The amino acid sequences of the SRO proteins were aligned with ClustalX, and the phylogenetic tree was constructed using the neighbor joining method in MEGA 7.0 software (**a**). Each node is represented by a number that indicates the bootstrap value for 1000 replicates. The right side illustrates the exon-intron organization of the corresponding *SRO* genes. The exon and intron are represented by the yellow boxes and black lines, respectively. The scale bar represents 1 kb (**b**). The blue boxes represented upstream/downstream.

2.4. Conserved Motifs Analysis of BrSRO Proteins

The phylogenetic relationship and classification of *BrSROs* were further supported by motif analysis (Figure 4). Ten (10) conserved motifs of *BrSROs* were captured by motif analysis using MEME suite. All genes have the motifs (motif 1, 2, 3, 4, 5, 6, 7, and 9) in A group except for *BrSRO1*. The genes of the H group (*BrSRO7*, *BrSRO8*, and *BrSRO11*) have the same motifs, which are motif 1, 2, 4, 6, and 8. In addition, the genes in group I (*BrSRO2* and *BrSRO5*) have the same motifs, which are motif 1, 2, 4, and 8. There are only three motifs (motif 2, 9, and 10) in Group K. Interestingly, all 12 genes have motif 2, which indicates that motif 2 is relatively conservative.

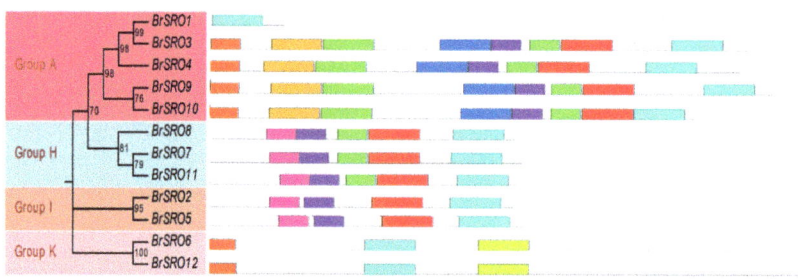

Figure 4. The conserved motifs of the BrSRO proteins based on the phylogenetic relationship. The BrSRO proteins phylogenetic relationship (**a**). The BrSRO proteins annotated with the MEME server (**b**). Distribution of the BrSRO conserved motifs in Chinese cabbage was analyzed by the online tool MEME. Ten motifs are marked by different colors.

2.5. Cis-Elements in the Promoters of BrSRO Genes

In order to study the response of *BrSRO* gene to various signal factors, we searched 2 kb sequences upstream of the start codon of *BrSROs* family for elements related to stress response. There are many light signal elements (MRE, box-4, TCT motif, etc.), hormone and stress response elements. The cis-acting elements related to hormones and stress response in *BrSRO* gene promoter were analyzed and illustrated. It can be seen from the table that *BrSRO* gene promoter contains 12 cis-elements that respond to hormones and stress. Interestingly, 12 *BrSRO* genes include at least one of the 12 predicted types of cis-elements in their promoter regions (Table 2). 10 *BrSRO* genes contain the ABRE cis-element; only two genes (*BrSRO1* and *BrSRO9*) lack it. There are more MeJA response elements (CGTCA-motif, TGACG-motif) in the *BrSRO* genes than other cis-elements. MBS is located in *BrSRO1, 3, 4, 7, 8, 11,* and *12*. All genes except *BrSRO1, 4, 5, 8,* and *10* have TATC-motif/P-box, indicating they are related to GA response. Only *BrSRO3, 4, 5,* and *7* have LTR and only two genes, namely, *BrSRO2* and *BrSRO8*, contain the TC-rich repeats cis-element in their promoter regions. These results suggest that SRO family may play a crucial role in the growth and development of Chinese cabbage, as well as in various hormones and stress.

Table 2. Putative cis-elements existed in the 2 kb upstream region of *BrSRO* gene family.

Gene	Hormonal Response Cis-Elements								Anaerobic Induction Response Element	Stress Response Cis-Elements		
	Abscisic Acid Response Element	Methyl jasmonate Response Element		Salicylic Acid Response Element	Auxin Response Element	Gibberellin Response Element				Drought Response Element	Low-Temperature Response Element	Defense and Stress Response Element
	ABRE	CGTCA-Motif	TGACG-Motif	TCA-Element	TGA-Element	GARE-Motif	TATC-Box	P-Box	ARE	MBS	LTR	TC-Rich Repeats
BrSRO1	0	3	3	0	0	1	0	0	2	3	0	0
BrSRO2	4	2	2	0	3	0	2	1	3	0	0	1
BrSRO3	1	0	0	1	0	0	1	1	3	4	1	0
BrSRO4	5	0	0	1	2	0	0	0	4	1	1	0
BrSRO5	5	0	0	0	2	0	0	0	4	1	1	0
BrSRO6	1	2	2	0	2	0	0	0	0	1	0	0
BrSRO7	2	0	0	0	1	0	0	1	5	0	1	0
BrSRO8	3	5	5	0	0	0	0	0	1	1	0	1
BrSRO9	0	3	3	0	1	0	0	0	2	2	0	0
BrSRO10	1	4	4	1	1	1	0	3	4	0	1	0
BrSRO11	3	3	3	0	0	0	0	1	0	0	0	0
BrSRO12	5	4	4	1	1	0	0	2	0	2	0	0

2.6. Relative Expression of 12 BrSRO Genes

Using qRT-PCR, the relative expression levels of *BrSRO* genes in leaf were analyzed under abiotic stresses for 24 h, 48 h, and 72 h. The results showed that the expression of *BrSROs* responded differently to various abiotic stresses. Under high temperature stress, the relative expression levels of *BrSRO1, 5, 6,* and *8* genes were up-regulated and the rest of genes was down-regulated at 24 h. The relative expression levels of *BrSRO1, 8,* and *9* genes were up-regulated and *BrSRO4* and *BrSRO5* were down-regulated at 48 h, while *BrSRO4, 5* and *BrSRO8* were up-regulated at 72 h. Interestingly, the relative expression level of *BrSRO8* was significantly up-regulated at three time points and reached the highest level at 24 h (Figure 5). Under low temperature, the relative expression levels of *BrSRO1, 3, 7, 8, 9,* and *12* genes were up-regulated at three time points and the up-regulated amplitudes of different genes were different at different time points (Figure 5). Under drought stress, the relative expression levels of *BrSRO1, 5,* and *9* genes were up-regulated at three points and the relative expression level of *BrSRO5* reached the highest level at 72 h while *BrSRO9* reached the highest level at 24 h and 48 h (Figure 5). Under 2%NaCl treatment, the relative expression level of all the *BrSRO* genes were up-regulated at 24 h, moreover, *BrSRO5* and *BrSRO8* were significantly up-regulated and about 7.5 times the control. At 48 h, the *BrSRO1, 3, 7, 8, 9,* and *11* were up-regulated while *BrSRO2, 3, 4,* and *5* were reached lowest for 48 h. Only *BrSRO1, 8* and *12* up-regulated at 24 h, 48 h, and 72 h (Figure 5). Thus, it could be seen that the up-regulation of *BrSRO1* and *8* genes were significant under all treatments, while the up-regulation of *BrSRO9* was significant under drought, low temperature, and salt stresses. The expression of *BrSRO12* was not significantly up-regulated or down-regulated in all treatments compared with the control. *BrSRO5* gene was significantly up-regulated under drought and salt treatments, and *BrSRO7* gene was significantly up-regulated under drought and low-temperature treatments. The above candidate genes (*BrSRO1, 5, 7, 8,* and *9*) were used for functional analyses in the succeeding experiment.

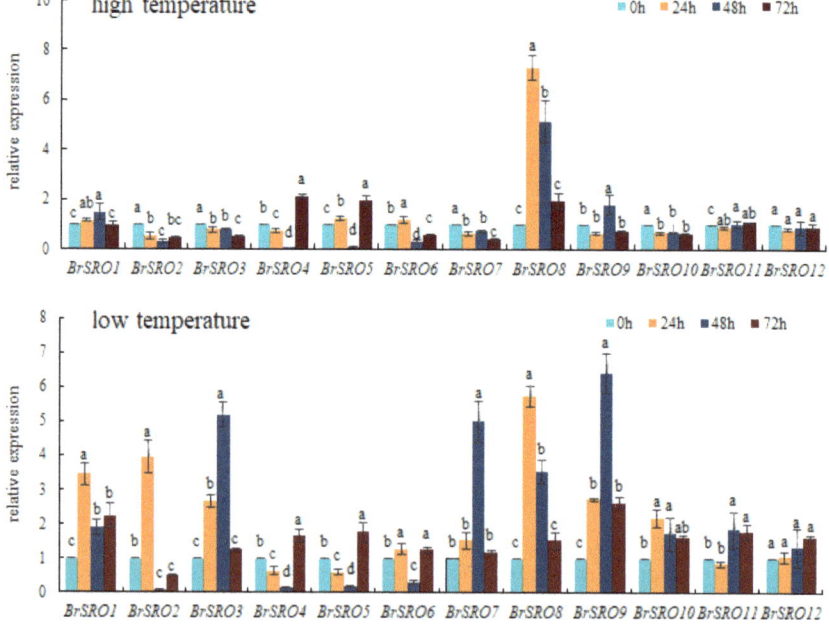

Figure 5. *Cont.*

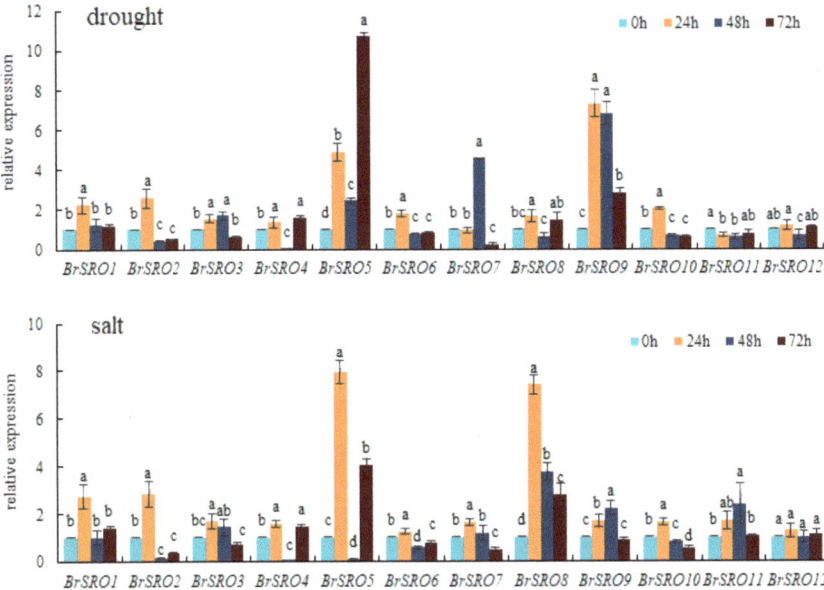

Figure 5. Expression profiles of 12 *BrSRO* genes in response to high temperature treatment, low-temperature treatment, drought treatment, and salt treatment. Quantitative reverse transcription polymerase chain reaction (qRT-PCR) analyses were used to assess the transcript levels of *BrSROs* in leaves sampled at 24 h, 48 h and 72 h after high temperature, low-temperature treatment, drought, and salt treatment in Chinese cabbage seedlings. 0 h as control. Three sets of repeats are set for each process. Error bars indicate standard deviations of three replicates and different letters describe significant differences at $p \leq 0.05$ level among different time points within the same gene.

3. Discussion

The SRO protein family is highly conserved and found in all land plant species [5]. Several *SROs* have been identified as involved in plant development and stresses response. However, the family members and functions of *SROs* are largely unknown in Chinese cabbage. The exact biochemical functions of the SRO proteins remain unknown. The *SROs* are characterized by the plant-specific domain architecture which contains a poly(ADP-ribose) polymerase catalytic (PARP, PF00644) and a C-terminal RCD1-SRO-TAF4 (RST, PF12174) domain [5]. In addition to these two domains, some SRO proteins have an N-terminal WWE domain (PF02825). The RST domain is plant-specific and present in SROs and TAF4 proteins. Previous studies have demonstrated that PARP-RST domains are specific to plants, while WWE-PARP domains are widely conserved in organisms even as distantly related as humans [20,21]. The RST domain is essential for the interaction between RCD1 and other TFs [3]. PARPs are a class of enzymes that are involved in many biological processes, including DNA damage repair, transcription, cell death pathways, and chromatin modification/remodeling [22]. In this study, a total of 12 *BrSROs* genes were identified from the Chinese cabbage genome and named *BrSRO1–BrSRO12*, according to chromosome location. Only two genes, *BrSRO4* and *BrSRO9*, were identified as having the WWE domain, whereas the rest of *BrSROs* only have the RST domain and PARP domain, lacking WWE domain. From the analysis of physicochemical properties of protein, the number and molecular weight of amino acids are quite different between *BrSRO1* and *BrSRO12*, which indicates that there are some differences in their structure and function. Phylogenetic tree analysis showed that SRO proteins of Chinese cabbage and *Arabidopsis thaliana* were highly similar, and their genetic relationships were also similar, and we can infer that there is functional similarity. The study of exons and introns is helpful to understand the differences of gene structure and function [23].

The number of *BrSRO* exons in the same group was very close, so most genes showed conservative gene structure, which supported a close evolutionary relationship [23]. Interestingly, the K group was located in the exon-rich region. It was proposed that the rates of intron creation are higher during earlier periods of plant evolution [24]. Additionally, the rate of intron loss is greater than the rate of intron gain after segmental duplication. Thus, it is possible that the group K may represent the original genes of *SRO* family [24]. Motif analysis further demonstrated the structural similarity of A, H, I, and K groups. All genes have motif 2, indicating that RST domain exists in motif 2. All genes contain cis-acting elements of light response. There are more gibberellin response elements, and methyl jasmonate response elements. Methyl jasmonate elements are important phytohormones that mediate plant development and defense mechanisms against biotic (i.e., necrotrophic pathogen infection and herbivorous insect attack) and abiotic (i.e., mechanical wounding) stress [25]. Methyl jasmonate can also be used as the core signal factor of plant resistance to insect invasion [26]. Salicylic acid (SA) is an important signaling molecule for plants to cope with biotic or abiotic stress [12]. Gibberellins are phytohormones that regulate multiple developmental processes, such as seed germination, stem elongation, flowering, and fruit development [27]. Many cis elements related to abscisic acid and drought stress response were found in the promoter region of *BrSRO* gene, which indicated that *BrSRO* gene family might respond to drought stress through the hormone signal transduction pathway.

The *SRO* family not only affects plants growth and development, but also affects their response to various stresses. The *SRO* family has proven to be able to respond to abiotic stress in many plants. For example, the relative expression level of *OsSRO1c* was significantly up-regulated under ABA and JA treatments. *Ta-SRO1* can regulate the oxygen content in wheat. Chemical reduction balance was used to improve the tolerance to drought, high salt, and H_2O_2 stress [1]. The expression level of *MdSRO4* in apples treated with 100 μmol L^{-1} ABA and 4 °C were 14 and 37 times higher than that of ABA and 4 °C, respectively. Under 20% polyethylene glycol (PEG) treatment, the relative expression levels of *MdRCD1*, *MdSRO2*, and *MdSRO3* were up-regulated by 18, 17, and 14 times compared with that of *MdSRO4*, respectively, indicating that *MdSRO4* could respond to ABA and chilling stress, *MdRCD1*, *MdSRO2*, and *MdSRO3* could respond to drought stress [13]. In this study, the expression levels of *BrSRO* genes in leaves were analyzed under abiotic stresses for 24 h, 48 h, and 72 h. Our results showed that the responses of *BrSROs* were different among heat, low temperature, drought, and salt stresses. *BrSRO8* is sensitive to high temperature, and the expression of *BrSRO1*, *3*, *7*, *8*, and *9* was higher under low temperature treatment. The response to drought stress was *BrSRO1*, *5*, and *9*, and to NaCl stress was *BrSRO7* and *BrSRO8*. The expression level changed with the time of treatment; it may be that plants regulate themselves to resist changes in the external environment. Interestingly, the expression levels of all 12 genes were up-regulated after 24 h salt treatment, indicating all the genes responded to salt stress at 24 h. Excess salts in soils cause growth arrest, molecular damage, and even the death of many of the salt-sensitive crop species that are grown today [28,29]. Thus it can be predicted that the *BrSRO* genes family may play an important role in the resistance to salt stress. Two *SRO* genes, *GHSRO04* (Gen bank accession number kr534896) and *GHSRO08* (Gen bank accession number kr534895) have been cloned from upland cotton. The two genes were induced to express by high salt and drought, indicating that *SRO* plays an important role in regulating the growth and development of cotton under pathogen attack, salt, and drought stresses, and has potential utilization value for the genetic improvement of cotton germplasm [30]. Whether the function of *SRO* genes in Chinese cabbage work under biotic and abiotic stresses or not, the candidate genes with higher expression levels (*BrSRO1*, *5*, *7*, *8*, and *9*) at three time points under abiotic stresses were selected to further verify their functions in the future.

4. Materials and Methods

4.1. Identification and Sequence Analysis of SRO Genes in Chinese Cabbage

Six known ID of *Arabidopsis thaliana SRO* genes [12] were put into *Arabidopsis* genome database (TAIR) [31] to obtain their protein sequences. Using *Arabidopsis* SRO protein sequences as probes, the candidate members of Chinese cabbage SRO family were searched and the coding sequences (CDS) and amino acid sequences of the *B. rapa* SRO genes were downloaded from the Brassica database [32]. The banana SRO genes and protein sequences were downloaded from the Banana Genome Hub [33], the rice SRO genes and protein sequences from the Rice Genome Annotation Project [34], and the website of Phytozome [35] was used to search for the SROs from *Solanum lycopersicum* and *Zea mays*. The candidate sequences with conservative domains of PARP (PS51059) and RST (PF12174) were then inspected using the SMART program [36]. Subsequently, the Prot-Param tool [37] was used to analyze the physicochemical parameters (i.e., length, molecular weight, and isoelectric point) of the SRO proteins. Subcellular localization prediction was carried out with the Plant-mPLoc [38].

4.2. Phylogenetic Analysis of SRO Genes in Chinese Cabbage

The phylogenetic tree was constructed with MEGA 7.0 (https://www.megasoftware.net/home) [39] on the basis of alignment with the amino acid sequences of the *BrSRO* proteins using the neighbor-joining method [40] with 1000 bootstrap replicates [41].

4.3. Gene Structure and Conserved Motifs Analysis of BrSROs

The distribution of the conserved motifs based on amino acid sequence was conducted with the online MEME program [42] and the MEME search was carried out with the following parameters: maximum number of motifs set at 10, a minimum width of 6 and a maximum width of 50. The other parameters were set as default. The exon-intron structure of each *BrSRO* was determined by aligning the full-length cDNA sequence with the genomic DNA sequence. The schematic structure of each *BrSRO* was constructed using the Gene Structure Display Server (GSDS 2.0) (http://gsds.cbi.pku.edu.cn) [43].

4.4. Chromosomal Distribution and Cis-Element Analyses of SRO Genes in Chinese Cabbage

The information about chromosomal distribution was obtained from the Chinese cabbage genome database [32], and the chromosomal location of *BrSRO* genes was illustrated from top to bottom concerning their position in the genome annotation using Mapchart [44]. For identification of cis-elements located at the promoter regions of *SRO* genes, the 2000 bp genomic DNA sequences upstream before the initiation codon (ATG) of each *BrSRO* gene were downloaded from the Chinese cabbage genome database. The PlantCARE database [45] was utilized to search the cis-regulatory elements in promoter regions of Chinese cabbage genes.

4.5. Plant Materials, Growth Conditions and Treatments

In this study, the plants used for expression analysis were sampled from the "furui" Chinese cabbage seedlings. The seeds were soaked in water for 2 h then placed on moist filter paper in petri dish, and finally kept in the dark to germinate at 25 °C for 16 h. After germination, uniformly geminated seeds were sown in 50-hole tray filled with substrate and then put in an artificial climate chamber. The growth condition of the artificial climate chamber was as follows: photoperiod 12 h/12 h, temperature 25 °C/18 °C (day/night), relative humidity 80%, light intensity 250 $\mu mol \cdot m^{-2} \cdot s^{-1}$. After sowing for 23 days, the uniform seedlings were selected and treated with low temperature 10 °C/5 °C (day/night), high temperature 35 °C/20 °C (day/night), 2% NaCl solution, and under natural drought conditions for 0 h, 24 h, 48 h, and 72 h. The leaves treated with high temperature, low temperature, salt and drought stress for 0 h, 24 h, 48 h and 72 h were sampled, which was frozen with liquid nitrogen and stored at −80 °C for the following experiment.

4.6. RNA Isolation and qRT-PCR Analysis

Total RNA was extracted from leaf tissues by using the Plant RNA Extraction Kit (Takara, Kusatsu, Japan). The first-strand cDNA fragment was synthesized from total RNA by using the Prime Script® RT Reagent kit (Takara, Kusatsu, Japan). The reverse transcripts were preserved at 20 °C for the following PCR amplification. The CDS sequences of *BrSRO* genes were input into the homepage of Shanghai biology company (Shanghai, China) for online primer design (as shown in Table 3), and then the primer sequences were synthesized. The *actin* gene was used for internal reference. The amplification system contained 2 µL cDNA, upstream primers 0.6 µL, downstream primers 0.6 µL, Rox 0.4 µL, SYBR 10 µL, reaction mix 6.4 µL, and ddH$_2$O 20 µL. The PCR cycling conditions included an initial polymerase activation step of 95 °C for 15 min, followed by 40 cycles of 95 °C for 10 s, and 60 °C for 30 s. Three biological replications for each sample were done. The relative expression levels of the *BrSRO* gene are represented in the form of relative changes by the $2^{-\Delta\Delta Ct}$ method [46]. Three biological replicates were carried out and the significance was determined with SPSS software. (SPSS 17.0, IBM, Chicago, IL, USA) ($p \leq 0.05$)

Table 3. The sequences of primers used for qRT-PCR.

Gene Name	Forward Primer Sequence (5'-3')	Reverse Primer Sequence (5'-3')
BrSRO01	AAGCTGAGGATGATTGTTGGAGA	CAAAGCAGTGTGTGGTAAGCG
BrSRO02	GGGTTTGCCGCCGTTGGATC	TTTGCCGCCGCCTTCTTCAC
BrSRO03	AAGCCTGCTGAGGAGGAAGACC	CGACGCCACCTGAAAACCTATACG
BrSRO04	GAACTCACGGCTCACCTTGGAAG	GAGCAGAGGGTAAGGCATCAAAGC
BrSRO05	AGCTGCGGAGTCGGAAGATGG	CCTCGTGGAACAACCTCAGACTTC
BrSRO06	AATGAATGCTCGTGGTCCGTTGG	GCTTGGTGGTGGCGGTGAAG
BrSRO07	GCGATCACCACGAGAGCCAAG	AGCCAGCGTACCAACCGTATTTG
BrSRO08	GCGGAGGCTATGAAGAGGAAGAAC	CGACCTCGCTGCTGCTAAACC
BrSRO09	CACCAAACCCGCAGACCCAAG	TGACCAGCGACTTCCCAGAGC
BrSRO10	TCTGGTGTCAAGCCTGCTGGAG	CGAGCTTCCGCAATCTCACTGG
BrSRO11	GCGGTTGTGTCAGTGCTGTCC	GCCACTTGTCTCATCTTCCGAACC
BrSRO12	GTGTGGAAGAAAGGATGCGAGGAC	CGTTGATTTGCTGCCGAACATCTG
actin	CCAGGAATCGCTGACCGTAT	CTGTTGGAAAGTGCTGAGGGA

Author Contributions: Conceptualization, J.Y. and L.H.; methodology, Y.Q., X.G., Z.L. and Y.W.; software, Y.Q., X.G.; validation, Y.Q. and X.G.; formal analysis, Y.Q.; investigation, Y.Q. and X.G.; resources, Y.Q.; supervision, J.Y.; data curation, Y.Q. and L.H.; writing—original draft preparation, Y.Q.; writing—review and editing, Y.Q., X.G., Z.L., Y.W., J.Y. and L.H.; project administration, J.Y. and L.H.; funding acquisition, J.Y. and L.H. All authors have read and agreed to the published version of the manuscript.

Funding: This research was supported by National Natural Science Foundation of China (31660584), National Key R&D Plan special plan (17ZD2NA015), Gansu Provincial Natural Science Foundation (18JR3RA166), Scientific Research Start-up Funds for Openly-Recruited Doctors (GSAU-RCZX201713) and Sheng Tongsheng Innovation Funds (GSAU-STS-2018-32).

Conflicts of Interest: The authors declare no conflict of interest.

References

1. Liu, S.; Liu, S.; Wang, M.; Wei, T.; Meng, C.; Wang, M.; Xia, G. A wheat similar to rcd-one gene enhances seedling growth and abiotic stress resistance by modulating redox homeostasis and maintaining genomic integrity. *Plant Cell* **2014**, *26*, 164–180. [CrossRef] [PubMed]
2. You, J.; Zong, W.; Du, H.; Hu, H.; Xiong, L. A special member of the rice sro family, ossro1c, mediates responses to multiple abiotic stresses through interaction with various transcription factors. *Plant Mol. Biol.* **2014**, *84*, 693–705. [CrossRef] [PubMed]
3. Jaspers, P.; Blomster, T.; Brosché, M.; Salojärvi, J.; Ahlfors, R.; Vainonen, J.P.; Reddy, R.A.; Immink, R.; Angenent, G.; Turck, F.; et al. Unequally redundant rcd1 and sro1 mediate stress and developmental responses and interact with transcription factors. *Plant J.* **2009**, *60*, 268–279. [CrossRef] [PubMed]

4. Citarelli, M.; Teotia, S.; Lamb, R.S. Evolutionary history of the poly(adp-ribose) polymerase gene family in eukaryotes. *BMC Evol. Biol.* **2010**, *10*, 308. [CrossRef]
5. Jaspers, P.; Overmyer, K.; Wrzaczek, M.; Vainonen, J.P.; Blomster, T.; Salojärvi, J.; Reddy, R.A.; Kangasjärvi, J. The rst and parp-like domain containing sro protein family: Analysis of protein structure, function and conservation in land plants. *BMC Genom.* **2010**, *11*, 170. [CrossRef]
6. Katiyar-Agarwal, S.; Zhu, J.; Kim, K.; Agarwal, M.; Fu, X.; Huang, A.; Zhu, J.K. The plasma membrane na+/h+ antiporter sos1 interacts with rcd1 and functions in oxidative stress tolerance in *Arabidopsis*. *Proc. Natl. Acad. Sci. USA* **2006**, *103*, 18816–18821. [CrossRef]
7. Ahlfors, R.; Lång, S.; Overmyer, K.; Jaspers, P.; Brosché, M.; Tauriainen, A.; Kollist, H.; Tuominen, H.; Belles-Boix, E.; Piippo, M.; et al. *Arabidopsis* radical-induced cell death1 belongs to the wwe protein-protein interaction domain protein family and modulates abscisic acid, ethylene, and methyl jasmonate responses. *Plant Cell* **2004**, *16*, 1925–1937. [CrossRef]
8. Vainonen, J.P.; Jaspers, P.; Wrzaczek, M.; Lamminmäki, A.; Reddy, R.A.; Vaahtera, L.; Brosché, M.; Kangasjärvi, J. Rcd1-dreb2a interaction in leaf senescence and stress responses in *Arabidopsis thaliana*. *Biochem. J.* **2012**, *442*, 573–581. [CrossRef]
9. Teotia, S.; Lamb, R.S. The paralogous genes radical-induced cell death1 and similar to rcd one1 have partially redundant functions during *Arabidopsis* development. *Plant Physiol.* **2009**, *151*, 180–198. [CrossRef] [PubMed]
10. Zhao, X.; Gao, L.; Jin, P.; Cui, L. The similar to rcd-one 1 protein sro1 interacts with gpx3 and functions in plant tolerance of mercury stress. *Biosci. Biotechnol. Biochem.* **2018**, *82*, 74–80. [CrossRef]
11. Borsani, O.; Zhu, J.; Verslues, P.E.; Sunkar, R.; Zhu, J.K. Endogenous sirnas derived from a pair of natural cis-antisense transcripts regulate salt tolerance in *Arabidopsis*. *Cell* **2005**, *123*, 1279–1291. [CrossRef] [PubMed]
12. Li, B.Z.; Zhao, X.; Zhao, X.L.; Peng, L. Structure and function analysis of *Arabidopsis thaliana* sro protein family. *Yi Chuan Hered.* **2013**, *35*, 1189–1197. [CrossRef] [PubMed]
13. Li, H.; Li, R.; Qu, F.; Yao, J.; Hao, Y.; Wang, X.; You, C. Identification of the sro gene family in apples (malus × domestica) with a functional characterization of mdrcd1. *Tree Genet. Genomes* **2017**, *13*, 94. [CrossRef]
14. You, J.; Zong, W.; Li, X.; Ning, J.; Hu, H.; Li, X.; Xiao, J.; Xiong, L. The snac1-targeted gene ossro1c modulates stomatal closure and oxidative stress tolerance by regulating hydrogen peroxide in rice. *J. Exp. Bot.* **2013**, *64*, 569–583. [CrossRef]
15. Cheng, F.; Liu, S.; Wu, J.; Fang, L.; Sun, S.; Liu, B.; Li, P.; Hua, W.; Wang, X. Brad, the genetics and genomics database for brassica plants. *BMC Plant Biol.* **2011**, *11*, 136. [CrossRef]
16. Huang, X.Y.; Tao, P.; Li, B.Y.; Wang, W.H.; Yue, Z.C.; Lei, J.L.; Zhong, X.M. Genome-wide identification, classification, and analysis of heat shock transcription factor family in chinese cabbage (*Brassica rapa* pekinensis). *Genet. Mol. Res. GMR* **2015**, *14*, 2189–2204. [CrossRef]
17. Tao, P.; Zhong, X.; Li, B.; Wang, W.; Yue, Z.; Lei, J.; Guo, W.; Huang, X. Genome-wide identification and characterization of aquaporin genes (aqps) in chinese cabbage (*Brassica rapa* ssp. Pekinensis). *Mol. Genet. Genom. MGG* **2014**, *289*, 1131–1145. [CrossRef]
18. Liu, Y.; Guan, X.; Liu, S.; Yang, M.; Ren, J.; Guo, M.; Huang, Z.; Zhang, Y. Genome-wide identification and analysis of tcp transcription factors involved in the formation of leafy head in chinese cabbage. *Int. J. Mol. Sci.* **2018**, *19*, 847. [CrossRef]
19. Wang, Z.; Tang, J.; Hu, R.; Wu, P.; Hou, X.L.; Song, X.M.; Xiong, A.S. Genome-wide analysis of the r2r3-myb transcription factor genes in chinese cabbage (*Brassica rapa* ssp. Pekinensis) reveals their stress and hormone responsive patterns. *BMC Genom.* **2015**, *16*, 17. [CrossRef]
20. Otto, H.; Reche, P.A.; Bazan, F.; Dittmar, K.; Haag, F.; Koch-Nolte, F. In silico characterization of the family of parp-like poly(adp-ribosyl)transferases (parts). *BMC Genom.* **2005**, *6*, 139. [CrossRef]
21. Hassa, P.O.; Hottiger, M.O. The diverse biological roles of mammalian parps, a small but powerful family of poly-adp-ribose polymerases. *Front. Biosci. J. Virtual Libr.* **2008**, *13*, 3046–3082. [CrossRef] [PubMed]
22. Kim, M.Y.; Zhang, T.; Kraus, W.L. Poly(adp-ribosyl)ation by parp-1: 'Par-laying' nad+ into a nuclear signal. *Genes Dev.* **2005**, *19*, 1951–1967. [CrossRef] [PubMed]
23. Xu, G.; Guo, C.; Shan, H.; Kong, H. Divergence of duplicate genes in exon-intron structure. *Proc. Natl. Acad. Sci. USA* **2012**, *109*, 1187–1192. [CrossRef] [PubMed]
24. Roy, S.W.; Penny, D. Patterns of intron loss and gain in plants: Intron loss-dominated evolution and genome-wide comparison of o. Sativa and a. Thaliana. *Mol. Biol. Evol.* **2007**, *24*, 171–181. [CrossRef] [PubMed]

25. Cheong, J.J.; Choi, Y.D. Methyl jasmonate as a vital substance in plants. *Trends Genet. TIG* **2003**, *19*, 409–413. [CrossRef]
26. Chini, A.; Fonseca, S.; Fernández, G.; Adie, B.; Chico, J.M.; Lorenzo, O.; García-Casado, G.; López-Vidriero, I.; Lozano, F.M.; Ponce, M.R.; et al. The jaz family of repressors is the missing link in jasmonate signalling. *Nature* **2007**, *448*, 666–671. [CrossRef]
27. Eriksson, S.; Böhlenius, H.; Moritz, T.; Nilsson, O. Ga4 is the active gibberellin in the regulation of leafy transcription and *Arabidopsis* floral initiation. *Plant Cell* **2006**, *18*, 2172–2181. [CrossRef]
28. Liang, W.; Ma, X.; Wan, P.; Liu, L. Plant salt-tolerance mechanism: A review. *Biochem. Biophys. Res. Commun.* **2018**, *495*, 286–291. [CrossRef]
29. Wang, Q.; Guan, C.; Wang, P.; Ma, Q.; Bao, A.K.; Zhang, J.L.; Wang, S.M. The effect of *Athkt1*;1 or *AtSOS1* mutation on the expressions of Na^+ or K^+ transporter genes and ion homeostasis in *Arabidopsis thaliana* under salt stress. *Int. J. Mol. Sci.* **2019**, *20*, 1085. [CrossRef]
30. Ahuja, I.; de Vos, R.C.; Bones, A.M.; Hall, R.D. Plant molecular stress responses face climate change. *Trends Plant Sci.* **2010**, *15*, 664–674. [CrossRef]
31. TAIR. Available online: http://www.Arabidopsis.org (accessed on 11 November 2015).
32. Wang, X.; Wang, H.; Wang, J.; Sun, R.; Wu, J.; Liu, S.; Bai, Y.; Mun, J.H.; Bancroft, I.; Cheng, F.; et al. The genome of the mesopolyploid crop species brassica rapa. *Nat. Genet.* **2011**, *43*, 1035–1039. [CrossRef] [PubMed]
33. Banana Genome Hub. Available online: http://banana-genome-hub.southgreen.fr/ (accessed on 17 September 2020).
34. RGAP. Available online: http://rice.plantbiology.msu.edu (accessed on 11 November 2015).
35. Phytozome. Available online: https://phytozome.jgi.doe.gov/pz/portal.html (accessed on 17 September 2020).
36. SMART. Available online: http://smart.embl.de/ (accessed on 11 November 2015).
37. Artimo, P.; Jonnalagedda, M.; Arnold, K.; Baratin, D.; Csardi, G.; de Castro, E.; Duvaud, S.; Flegel, V.; Fortier, A.; Gasteiger, E.; et al. Expasy: Sib bioinformatics resource portal. *Nucleic Acids Res.* **2012**, *40*, W597–W603. [CrossRef] [PubMed]
38. Yu, C.S.; Lin, C.J.; Hwang, J.K. Predicting subcellular localization of proteins for gram-negative bacteria by support vector machines based on n-peptide compositions. *Protein Sci. A Publ. Protein Soc.* **2004**, *13*, 1402–1406. [CrossRef] [PubMed]
39. Kumar, S.; Stecher, G.; Tamura, K. Mega 7: Molecular evolutionary genetics analysis version 7.0 for bigger datasets. *Mol. Biol. Evol.* **2016**, *33*, 1870–1874. [CrossRef] [PubMed]
40. Saitou, N.; Nei, M. The neighbor-joining method: A new method for reconstructing phylogenetic trees. *Mol. Biol. Evol.* **1987**, *4*, 406–425.
41. Felsenstein, J. Confidence limits on phylogenies: An approach using the bootstrap. *Evol. Int. J. Org. Evol.* **1985**, *39*, 783–791. [CrossRef]
42. Bailey, T.L.; Boden, M.; Buske, F.A.; Frith, M.; Grant, C.E.; Clementi, L.; Ren, J.; Li, W.W.; Noble, W.S. Meme suite: Tools for motif discovery and searching. *Nucleic Acids Res.* **2009**, *37*, W202–W208. [CrossRef]
43. Hu, B.; Jin, J.; Guo, A.Y.; Zhang, H.; Luo, J.; Gao, G. Gsds 2.0: An upgraded gene feature visualization server. *Bioinformatics (Oxford, England)* **2015**, *31*, 1296–1297. [CrossRef]
44. Voorrips, R.E. Mapchart: Software for the graphical presentation of linkage maps and qtls. *J. Hered.* **2002**, *93*, 77–78. [CrossRef]
45. Lescot, M.; Déhais, P.; Thijs, G.; Marchal, K.; Moreau, Y.; Van de Peer, Y.; Rouzé, P.; Rombauts, S. Plantcare, a database of plant cis-acting regulatory elements and a portal to tools for in silico analysis of promoter sequences. *Nucleic Acids Res.* **2002**, *30*, 325–327. [CrossRef]
46. Livak, K.J.; Schmittgen, T.D. Analysis of relative gene expression data using real-time quantitative pcr and the 2(-delta delta c(t)) method. *Methods (San Diego, CA)* **2001**, *25*, 402–408. [CrossRef] [PubMed]

© 2020 by the authors. Licensee MDPI, Basel, Switzerland. This article is an open access article distributed under the terms and conditions of the Creative Commons Attribution (CC BY) license (http://creativecommons.org/licenses/by/4.0/).

Article

Gibberellin Promotes Bolting and Flowering via the Floral Integrators *RsFT* and *RsSOC1-1* under Marginal Vernalization in Radish

Haemyeong Jung [1,2], Seung Hee Jo [1,2], Won Yong Jung [1], Hyun Ji Park [1], Areum Lee [1,2], Jae Sun Moon [1], So Yoon Seong [3], Ju-Kon Kim [3,4], Youn-Sung Kim [5,*] and Hye Sun Cho [1,2,*]

1. Plant Systems Engineering Research Center, Korea Research Institute of Bioscience and Biotechnology, Daejeon 34141, Korea; hmjung@kribb.re.kr (H.J.); chohee0720@kribb.re.kr (S.H.J.); jwy95@kribb.re.kr (W.Y.J.); hjpark@kribb.re.kr (H.J.P.); lar1027@kribb.re.kr (A.L.); jsmoon@kribb.re.kr (J.S.M.)
2. Department of Biosystems and Bioengineering, KRIBB School of Biotechnology, Korea University of Science and Technology, Daejeon 34113, Korea
3. Crop Biotechnology Institute/GreenBio Science and Technology, Seoul National University, Pyeongchang 25354, Korea; syseong7@snu.ac.kr (S.Y.S.); jukon@snu.ac.kr (J.-K.K.)
4. Graduate School of International Agricultural Technology, Seoul National University, Pyeongchang 25354, Korea
5. Department of Biotechnology, NongWoo Bio, Anseong 17558, Korea
* Correspondence: yskim0907@hanmail.net (Y.-S.K.); hscho@kribb.re.kr (H.S.C.); Tel.: +82-42-31-4323 (Y.-S.K.); +82-42-860-4469 (H.S.C.)

Received: 30 March 2020; Accepted: 29 April 2020; Published: 7 May 2020

Abstract: Gibberellic acid (GA) is one of the factors that promotes flowering in radish (*Raphanus Sativus* L.), although the mechanism mediating GA activation of flowering has not been determined. To identify this mechanism in radish, we compared the effects of GA treatment on late-flowering (NH-JS1) and early-flowering (NH-JS2) radish lines. GA treatment promoted flowering in both lines, but not without vernalization. NH-JS2 plants displayed greater bolting and flowering pathway responses to GA treatment than NH-JS1. This variation was not due to differences in GA sensitivity in the two lines. We performed RNA-seq analysis to investigate GA-mediated changes in gene expression profiles in the two radish lines. We identified 313 upregulated, differentially expressed genes (DEGs) and 207 downregulated DEGs in NH-JS2 relative to NH-JS1 in response to GA. Of these, 21 and 8 genes were identified as flowering time and GA-responsive genes, respectively. The results of RNA-seq and quantitative PCR (qPCR) analyses indicated that *RsFT* and *RsSOC1-1* expression levels increased after GA treatment in NH-JS2 plants but not in NH-JS1. These results identified the molecular mechanism underlying differences in the flowering-time genes of NH-JS1 and NH-JS2 after GA treatment under insufficient vernalization conditions.

Keywords: bolting; flowering time gene; gibberellin; radish (*Raphanus sativus* L.); RNA sequencing; *RsFT*; *RsSOC1*; vernalization

1. Introduction

Gibberellins are tetracyclic diterpene acids that are synthesized in plastids and then translocated into the cytosol in a biologically active form [1]. Bioactive gibberellic acids (GAs) control diverse processes throughout the plant life cycle, encompassing seed germination, stem and leaf growth, trichome development, flowering time [2], and vegetative and reproductive development [3]. GAs are involved in plant growth, development, cell expansion, and division, and respond to specific combinations of internal cues and external stimuli [4–6]. Several recent studies reported that GA signaling also is involved in abiotic stress adaptation along with other hormone-signaling pathways [7–9]. GAs are a large family of more than 130 structurally related compounds, although only a limited number of GAs display intrinsic biological activity [10]. GAs are synthesized at their sites of action to regulate growth, and GA levels are tightly regulated through a process of feedback regulation to maintain optimal levels for coordinating plant growth and development [11]. Our knowledge of the molecular mechanisms underlying GA signaling in plants was advanced by the following two discoveries: GIBBERELLIN-INSENSITIVE DWARF1 (GID1) encodes a soluble GA receptor, and the DELLA (Asp-Glu-Leu-Leu-Ala) transcriptional regulators negatively control the GA-signaling pathway [12,13]. GID1 binds bioactive GA in a deep binding pocket, and the N-terminal extension induces conformational changes that result in covering the GA pocket [12,13]. The GID1-GA complex can bind DELLA to produce the GID1-GA-DELLA protein complex, followed by the ubiquitin-ligase complex Skp, Cullin, F-box containing complex (SCF)-dependent degradation of DELLA protein, ultimately triggering GA-mediated downstream responses [14,15]. Manipulating endogenous GA levels is an established practice in agriculture to modulate plant stature, and the introduction of dwarf alleles into staple crops greatly increases grain yields [16,17]. For this reason, one of the leading aims in the Green Revolution was inducing semi-dwarfism traits, which improved harvest index, enhanced lodging resistance, and increased yield [18,19].

The induction of flowering is the most important event during the transition from the vegetative phase to the reproductive phase during the entire life cycle of higher plants. Flowering induction is precisely regulated by the interplay between endogenous cues and genetic pathways that respond to environmental stimuli such as photoperiod, vernalization, age, and autonomous and gibberellin pathways [20,21]. These signals converge on a small number of floral integrators, including FLOWERING LOCUS T (FT), SUPPRESSOR OF OVEREXPRESSION OF CONSTANS1 (SOC1), and LEAFY (LFY), eventually leading to the activation of floral-meristem identity genes [22]. GAs generally induce bolting and flowering in long-day and biennial plants, although GA is not a universal flowering stimulus [23]. Several studies reported that GAs had complex roles in flowering induction that varied under different circumstances and in different plant species. Under nonpermissive conditions, GAs inhibit flowering in perennial plants but not in long-day and biennial plants [24,25]. It is believed that GAs promote vegetative growth instead of reproductive growth, thereby inhibiting flowering [2]. Although GAs affect flowering in a species-dependent manner, their function in flower development is probably universal. GAs act directly or indirectly to upregulate flowering time gene expression in leaves, and FT protein moves as a mobile signal from the leaf to the shoot apex, where it activates SOC1 and LFY by repressing the DELLA negative transcriptional regulators [26]. In Arabidopsis, GA signaling is crucial for bolting and flowering regardless of the active photoperiod, which was verified in GA-deficient mutants that lacked the GA receptor [23]. GA has a crucial role in flowering under short-day conditions by activating SOC1 when vernalization is not sufficient to induce SOC1 activation, and evidence from GA signaling and biosynthesis mutants shows that SOC1 integrates a GA-dependent flowering pathway [27]. To deepen our understanding of the mechanism that orchestrates flowering in Arabidopsis, we must identify how GA, photoperiod, and vernalization interact. GAs only act under vernalized conditions in plants that require vernalization; nonvernalized plants were unable to bolt or flower in response to GA [28]. In this case, vernalization is a prerequisite for flowering; GA can compensate for photoperiod, but not for vernalization. Vernalization increases the endogenous GA content in Brassicaceae; however, the mechanism mediating GA-induced bolting and flowering has not been completely elucidated.

Radish (*Raphanus sativus* L.) is one of the most important root vegetable crops in the Brassicaceae family and is cultivated worldwide. The tap root contains minerals, vitamins, dietary flavonols, and high glucosinolate content; therefore, the crop's economic value is primarily determined by root characteristics. Premature bolting and flowering can cause poor root development and serious economic loss in radish crops, particularly in the spring. Agricultural productivity in radish depends on the avoidance of early bolting to produce high-quality leafy vegetables. Studies on the flowering pathways and molecular functions of flowering-related genes have progressed in recent years in model plants, but few studies have focused on radish. The reference radish genome has been sequenced [29,30]. The advent of next-generation sequencing technologies enables genome-wide gene expression profiling and large-scale discovery of flowering-related genes in radish under diverse biological conditions. Recent work investigated a putative model of the bolting and flowering regulatory networks in radish by performing comparative analyses of microRNA (miRNA)-differentially expressed gene (DEG) data from vegetative and reproductive leaves [31]. This study also identified 142 flowering time genes from several developmental tissues using de novo transcriptome analysis. Our group identified 218 radish flowering time genes and a large number of DEGs that responded to vernalization at different bolting times in the late flowering (NH-JS1) and early flowering (NH-JS2) inbred radish lines. We proposed that similar genes were expressed in the vernalization pathways of the two inbred radish lines, and the vernalization pathway was conserved between radish and Arabidopsis [32]. The comparative transcriptome results of the two radish inbred lines with different bolting times suggested a regulatory network of flowering time genes. Vernalization is a key process for bolting and flowering in radish; radish does not bolt or flower without vernalization even when plants are grown for more than 100 days after seed germination [33]. The vernalization threshold requirement is a prerequisite for GA activation of bolting and flowering in radish [34]. Although studies of radish flowering are currently in progress, there are no reports on the mechanism mediating the GA-induced transition from vegetative growth to reproductive growth in radish.

In this study, we performed RNA-seq and qPCR analyses to examine the effect of exogenous GA on global gene expression profiles in two radish inbred lines with different bolting times (NH-JS1 and NH-JS2), and identified the molecular mechanism regulating marginal vernalization.

2. Results

2.1. GA Effects on Bolting Time Significantly Differed between the Two Radish Inbred Lines

To identify the mechanism of GA-induced bolting and flowering in radish, we observed the effects of exogenous GA on bolting phenotypes in two inbred lines, NH-JS1 (late bolting) and NH-JS2 (early bolting). Seeds were subjected to 0, 0.1, and 10 mM GA during germination, and then the seedlings were vernalized for 0, 10, and 20 days to examine the bolting phenotypes under various vernalization durations. GA did not significantly affect the bolting time in NH-JS1 at 10 days of vernalization. By contrast, GA significantly affected bolting time in the early bolting NH-JS2 line (Figure 1A). In the absence of exogenous GA, neither inbred line bolted even when grown up to 10–50 days after vernalization (DAV) (data not shown). Low GA levels (0.1 mM) can compensate for insufficient vernalization to induce bolting in the NH-JS2 line, but not in the NH-JS1 line. These combined results indicate that NH-JS1 and NH-JS2 differ in their responses to GA and have different bolting characteristics. Statistical analysis of the bolting percentage (the number of bolting plants in each line, ($n = 20$)) is presented in Figure 1B. There were no bolting plants in either line in the absence of vernalization, even when subjected to high GA concentrations. Treatment with 10 days of vernalization and 10 mM GA can induce bolting at 25 DAV, but only in the early-bolting NH-JS2 line. Treatment with 10 days of vernalization and 0.1 or 10 mM GA followed by plant growth for 30–35 DAV resulted in 60% and >80% bolted plants, respectively, in NH-JS2, whereas NH-JS1 plants did not bolt within 30 DAV even with 10 mM GA. Subjecting plants to 20 days of vernalization in the absence of GA resulted in approximately 16% of NH-JS2 plants bolting at 35 DAV, whereas no NH-JS1 plants bolted. Longer vernalization duration enhanced the bolting percentage in both inbred lines in proportion to the GA concentration, although NH-JS1 was much less sensitive to GA than NH-JS2. For example, 0.1 mM GA resulted in 5-fold, 4-fold, and 2.5-fold higher bolting percentage in NH-JS2 than in NH-JS1 at 20, 25, and 30 DAV, respectively (Figure 1B). These results indicate that radish requires at least 20 days of vernalization to bolt, and GA can compensate or induce bolting and flowering under insufficient vernalization. We suggest that marginal vernalization is dominant, and GA modulates the vernalization response during the radish floral transition.

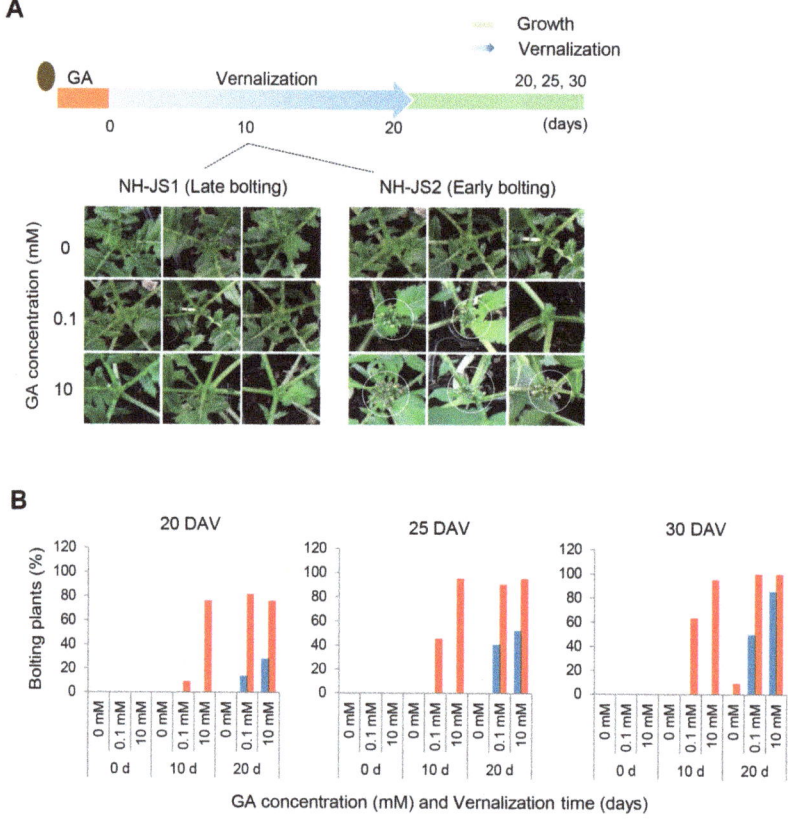

Figure 1. Phenotypes of NH-JS1 and NH-JS2 inbred radish lines under vernalization after gibberellin treatment. (**A**) Bolting phenotypes of NH-JS1 and NH-JS2 under vernalization for 10 days following gibberellin (GA) treatment (0, 0.1, or 10 mM) during seed germination. Seeds were germinated on filter paper in the presence or absence of GA and grown at 25 °C in a growth room. Germinated sprouts were vernalized by transferring into a cold room (5 ± 1 °C, 12 h light/12 h dark) for 0, 10, or 20 days. (**B**) Percentage of bolting radish plants after vernalization (n = 20 plants). DAV, days after vernalization.

2.2. GA Regulation Prioritizes the Floral Transition over Vegetative Growth in Radish

To determine whether GA hyposensitivity in NH-JS1 plants is specific for the bolting trait, we tested the effect of GA on hypocotyl length during seed germination. Figure 2 shows that the phenotypes of GA-treated seedlings were similar in both inbred lines, with growth proportionally enhanced relative to the GA concentrations (Figure 2A). The hypocotyl lengths were measured during a six-day period in the two inbred lines under different GA concentrations. In the presence of GA, hypocotyl lengths in both inbred lines increased gradually with increasing GA concentrations, and were nearly identical in both lines under each condition (Figure 2B). The hypocotyl length was slightly more sensitive to GA in NH-JS1 seedlings than in NH-JS2 at six days after 10 mM GA treatment (NH-JS1, 1.9-fold; NH-JS2, 1.6-fold) (Figure 2B). These combined results in both inbred lines indicate that GA effects on bolting phenotype depend on the duration of vernalization, but not on plant growth.

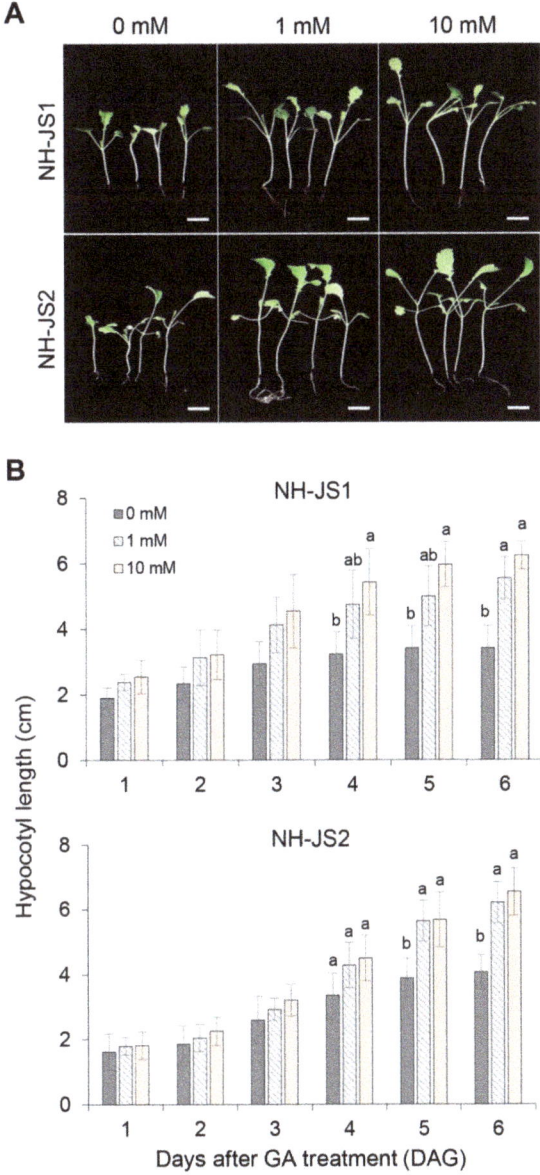

Figure 2. Hypocotyl phenotypes of the two inbred radish lines after GA treatment. (**A**) Hypocotyl phenotypes of the two radish lines after GA (0, 1, or 10 mM) treatment. Seeds were grown in a dark cold room at 5 ± 1 °C for 3 days and then transferred into a 25 °C growth room. Radish sprouts with hypocotyls longer than 0.5 cm were selected, and GA solution was poured directly over the seedlings (n = 6). Hypocotyl phenotypes were observed at 6 days after GA treatment. Scale bars = 0.5 cm. (**B**) Average hypocotyl lengths of NH-JS1 and NH-JS2. Bars represent the average hypocotyl lengths at different GA concentrations. Data are presented as mean ± SD (n = 6, except for 3–4 maximum and minimum seedlings). Asterisks mark significant differences between GA treatment compared with the absence of exogenous GA (one-way ANOVA followed by Bonferroni post hoc test).

To identify the precise effect of GA on radish bolting, we observed the GA-induced bolting time phenotype in the two inbred lines after minimal vernalization treatment or no vernalization (control). For this study, 4-day-old seedlings were vernalized for 10 days and acclimatized in a growth room for 14 days. Then, we directly sprayed 10 mM GA on the seedlings and compared the bolting phenotypes in the two lines at 17 days after GA treatment (DAG) (Figure 3A). In the absence of vernalization, none of the plants in the two lines bolted regardless of GA treatment, although both lines exhibited enhanced vegetative growth in response to GA (Figure 3A, 0 days vernalization). A 10-day vernalization treatment combined with exogenous GA treatment produced different bolting phenotypes in the two lines. All NH-JS2 plants bolted under these conditions, and NH-JS2 plants represent typical bolting and flowering characteristics in radish. We suggest that GA efficiently induces the floral transition under the conditions of a required minimum vernalization pretreatment. By contrast, NH-JS1 plants were hyposensitive to exogenous GA in the induction of bolting and flowering, and they did not bolt under the same conditions (Figure 3A, 10 days of vernalization). The effects of GA on plant growth were observable in both inbred lines. Statistical analysis of the bolting percentages (the number of bolting plants in each line ($n = 20$)) in response to GA treatment after 10 days of vernalization are presented in Figure 3B. The bolting percentages in early-bolting NH-JS2 plants differed with and without GA; without GA, approximately 70% of plants bolted at 30 DAG, and then no more plants bolted until 40 DAG, whereas up to 100% of GA-treated plants bolted before 25 DAG. By contrast, approximately 20% of late-bolting NH-JS1 plants treated with GA bolted at 30–40 DAG (Figure 3B). These results suggest that NH-JS1 plants require longer vernalization than NH-JS2, and NH-JS1 plants are less sensitive to GA than NH-JS2. GA can induce the floral transition under insufficient vernalization, but GA cannot independently induce the floral transition without vernalization.

2.3. Transcriptome Sequencing Identifies Genes Responding to GA in Radish

We conducted RNA-seq analysis to identify the genes involved in GA-mediated flowering induction in radish. First, we performed qPCR analysis to examine the effective time point of GA application on the expression of major flowering time genes in samples harvested at 6 h, and at 2, 4, and 6 days after GA treatment: 4-day-old seedlings were vernalized for 10 days and transferred to the growth room for 2 weeks to acclimatize, then 10 mM GA solution was sprayed on the leaves. The highest difference of flowering time gene expression level between two lines was observed at 6 days after GA treatment such as *RsMAF2* and *RsSOC1* in accordance with their bolting traits (Supplementary Figure S1). Therefore, we isolated RNA from shoot tissues, as stated above, collected 6 days after treatment with or without GA to identify global gene expression changes in the two inbred lines in response to GA as shown in Supplementary Figure S2A. A total of eight samples were analyzed from the two inbred lines, with two biological replicates for each condition. We constructed cDNA libraries and sequenced the libraries using an Illumina HiSeq 2000 Sequencing System (Supplementary Figure S2B). A total of 347,399,408 paired-end reads (lengths up to 101 base pair) were produced from eight generated libraries. Raw reads were subjected to quality control, and adapter sequences and low-quality reads were excluded. Approximately 72% clean reads assured the following criteria: Quality score $Q > 20$ and minimum read length ≥ 25 bp. Ultimately, a total of 251,429,330 clean reads was obtained from the libraries of the two inbred lines selectively treated with or without GA, and the average length of clean reads was 80.06 bp (Supplementary Table S1). To verify the similarity between two replicates, normalization counts were used to plot pairs of replicate samples. All samples had good reproducibility between pairs, showing 0.97–0.98 values (Supplementary Figure S2C). To analyze the proportion of unigenes in the eight transcriptome libraries, all clean reads were mapped to the 71,188 unigene reference sets, which was 94.65% (237,984,258) of reads from the two inbred lines selectively treated with GA and mapped to the reference unigenes. Only approximately 5% of all reads were unmapped (Supplementary Table S2). These combined results indicate that all of the transcriptome sets retained a high proportion of unigenes and were adaptable for subsequent DEG and expression profiling analyses.

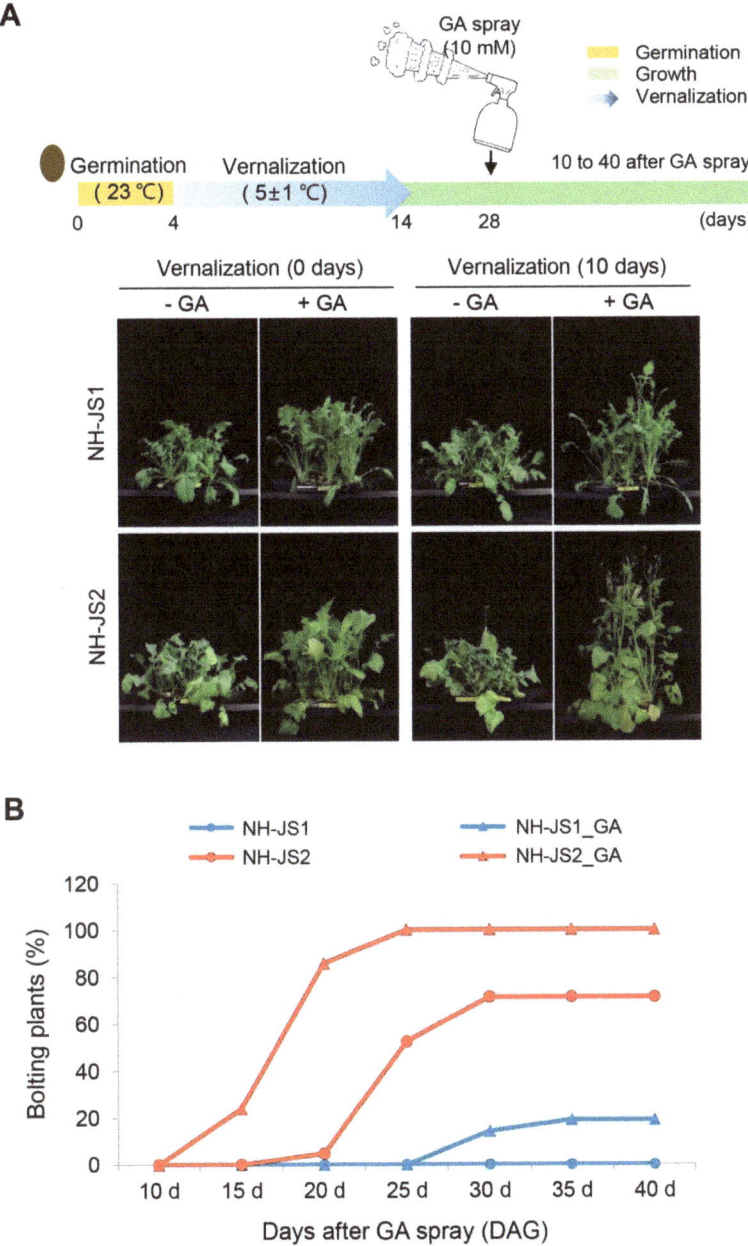

Figure 3. Phenotypes of the two inbred lines sprayed with gibberellin after vernalization. (**A**) Bolting phenotypes of the two inbred radish lines treated with GA (0 or 10 mM) after 10 days of vernalization. Four-day-old seedlings were vernalized for 10 days and then transferred into a 25 °C growth room for 2 weeks. Then, 10 mM GA solution was sprayed directly on the leaves (n = 20 plants). Plants were photographed at 17 days after GA spray. (**B**) Percentage of bolting radish plants after GA spray. The number of bolting plants was counted the next day after GA treatment. DAG, days after GA treatment.

2.4. Comparative Analysis of DEGs in the Two Inbred Radish Lines Treated with GA under Marginal Vernalization

To identify the genes involved in regulating GA-induced floral transition and bolting in radish, we performed a genome-wide comparative DEG analysis of the two inbred lines treated with GA. The total set of expressed genes was subjected to DEG analysis using the DESeq package in R. Then, the gene set was analyzed along with individual characteristics of the inbred line transcriptomes treated with GA. DEGs were analyzed using the following criteria: |log2 (fold-change)| ≥ 0.6, false discovery rate (FDR) ≤ 0.01, and read counts ≥ 500. Little flowering time- and GA-related DEGs were confirmed using |log2 (fold-change)| ≥ 1 conditions (data not shown); therefore, the fold-change was reduced to |log2 (fold-change)| ≥ 0.6. A total of 226 DEGs were obtained in each inbred line under GA treatment (GA was sprayed after 10 days of vernalization and 14 days of growth). Of these, 32 upregulated and 71 downregulated DEGs were detected in NH-JS1, and 55 upregulated and 86 downregulated DEGs were detected in NH-JS2 (Supplementary Figure S3A). Only 5 upregulated and 13 downregulated DEGs displayed overlapping regulation between the two inbred lines. The downregulated DEGs were more common than upregulated DEGs in both inbred lines. This result is consistent with the recent DEG analysis of *Rosa chinensis* treated with GA [35].

To identify the active biological pathways in each line in response to GA, we performed functional pathway enrichment analysis by comparing DEGs in NH-JS1_GA vs. NH-JS1 and NH-JS2_GA vs. NH-JS2 using the Kyoto Encyclopedia of Genes and Genomics (KEGG) database. A total of 83 DEGs from NH-JS1 and 103 DEGs from NH-JS2 did not show a significant distribution of specific pathways. We identified three KEGG pathways that contained the highest number of assigned DEGs in each line. Each group of upregulated DEGs were commonly assigned to 'plant hormone signal transduction' (NH-JS1 vs. NH-JS2, 7 vs. 4) and 'metabolic pathway' (7 vs. 3), whereas downregulated DEGs were commonly assigned to 'biosynthesis of secondary metabolites' (6 vs. 8) and 'metabolic pathways' (7 vs. 12). The 'plant-pathogen interaction' (0 vs. 7) and 'ribosome' (0 vs. 6) pathways were matchless pathways whose DEGs were upregulated and downregulated by GA, respectively, in NH-JS2 compared with NH-JS1. The 'galactose metabolism' (4 vs. 0) and 'zeatin biosynthesis' (2 vs. 0) pathways were unique to NH-JS1 as upregulated and downregulated DEGs, respectively (Supplementary Figure S3B).

2.5. Comprehensive Analysis of DEGs Related to GA-Responsive Flowering Pathways in Radish

We analyzed the differences in GA responses between the two inbred lines. A total of 3165 DEGs were identified in the two inbred lines with and without GA treatment. A total of 1324 upregulated and 960 downregulated DEGs were detected in NH-JS2 vs. NH-JS1 with GA treatment, although approximately 80% (1764 DEGs) overlapped with those in NH-JS2 vs. NH-JS1 without GA treatment. Consequently, 313 upregulated and 207 downregulated GA-specific DEGs were detected in NH-JS2 compared with NH-JS1 (Figure 4A). To further evaluate the GA effect on phenotypic variation between the two lines, we conducted statistical enrichment of GA-specific DEGs in the two lines using KEGG pathway analysis. The GA-specific DEGs in the two inbred lines (NH-JS2_GA vs. NH-JS1_GA) were enriched in 13 significant pathways (Supplementary Figure S4). The highly significant pathways included 'carotenoid biosynthesis', 'biosynthesis of secondary metabolites', 'glucosinolate biosynthesis', 'arachidonic acid metabolism', 'nitrogen metabolism', and 'ubiquinone and other terpenoid-quinone biosynthesis' ($p < 0.001$). In the absence of GA, the NH-JS2 vs. NH-JS1 DEGs primarily mapped to 'ribosome' followed by 'aminoacyl-transfer RNA biosynthesis', 'porphyrin and chlorophyll metabolism', and 'pantothenate and CoA biosynthesis' pathways. These combined results indicate that GA effects were indirectly rather than directly related to flowering or GA-related genes in the two inbred radish lines.

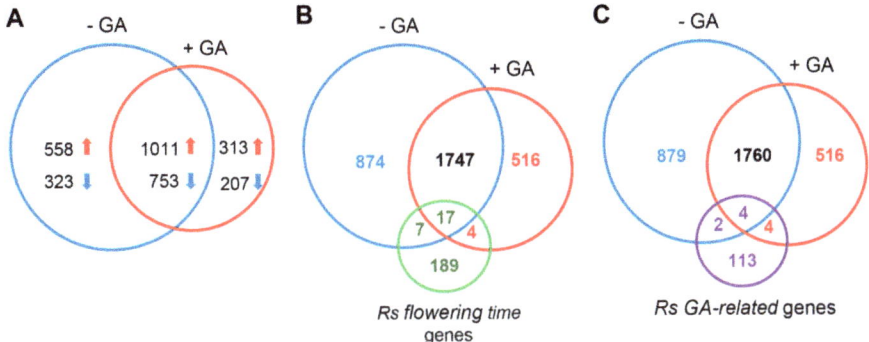
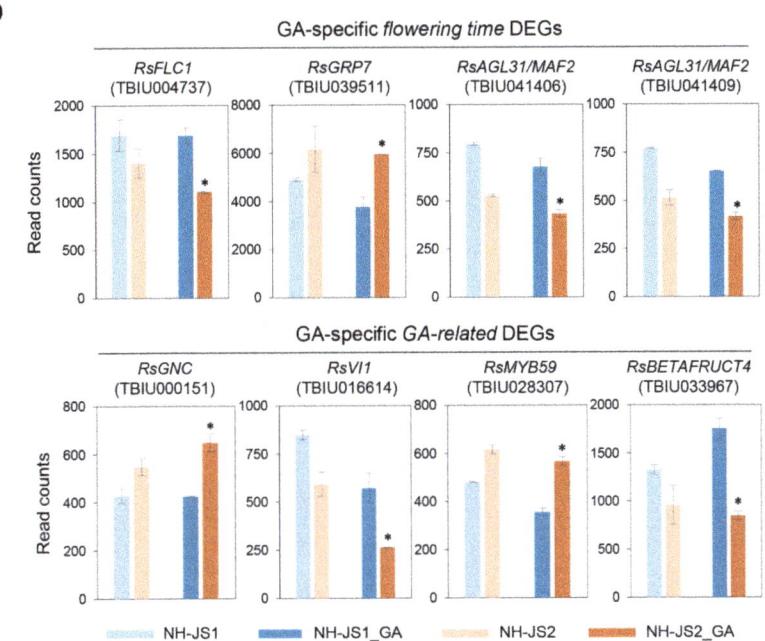

Figure 4. Comparative analysis of differentially expressed genes (DEGs) with and without gibberellin treatment in the two inbred radish lines ($|\log_2 FC \geq 0.6|$, false discovery rate (FDR) ≤ 0.01, read count ≥ 500). (**A**) Number of DEGs in the two lines with and without GA treatment. Red arrows, number of upregulated DEGs; blue arrows, number of downregulated DEGs. (**B**) The number of flowering time DEGs between two lines under GA treatment. Red, the number of flowering time DEGs after GA treatment. (**C**) The number of GA-related DEGs between two lines under GA treatment. Red, the number of GA-related DEGs after GA treatment. (**D**) RNA-sequencing results of flowering time and GA-specific DEGs. Asterisk means selected DEGs response to GA.

To identify flowering time genes that respond to GA and affect bolting and flowering in our radish transcriptome data sets, we examined 218 flowering time genes [32] and 125 GA-related genes using the interactive flowering database FLOWerRing interactive database (FLOR-ID) and published literature on DEGs in flowering-related pathways (Supplementary Table S3). We applied the same following criteria: $|\log_2 \text{(fold-change)}| \geq 0.6$, false discovery rate (FDR) ≤ 0.01, and read counts ≥ 500. We detected 15 upregulated DEGs and 6 downregulated flowering time DEGs in response to GA in the two inbred lines

(Figure 4B, Supplementary Table S4). A total of 17 flowering time DEGs in response to GA overlapped with those in the absence of GA in the two lines. *RsELF3* and *RsSOC1 Ft* DEGs were the most strongly upregulated among the GA-responsive DEGs. By contrast, *RsFLC* (Theragen Bio Institute Unigene: TBIU004737) and *RsMAF2*, which are flowering repressors, were downregulated in NH-JS2 with similar expression levels in the absence of GA as in our previous report [32]. Although no significant differences were observed with and without GA treatment, the differential expression of these flowering time DEGs was slightly increased by GA (Supplementary Table S4). We detected 6 upregulated and 4 downregulated GA-related DEGs in the two inbred lines that were classified as 'response to GA' using the same criteria (Figure 4C, Supplementary Table S5). The GA-regulated flowering activator *RsGASA6* [36] was more strongly upregulated in NH-JS2 than in NH-JS1, whereas *RsEXPA1*, *RsPIF4*, and *RsMYB28* or *RsMYB29* were more strongly downregulated in NH-JS2 than in NH-JS1 with GA treatment. The expression of GA-related DEGs *RsGNC*, *RsVI1*, *RsMYB59*, and *RsBETAFRUCT4* was unique to NH-JS2 in response to GA. RNA-seq analysis indicated that only 8 flowering time and GA-related DEGs were differentially expressed in response to GA between the two lines (Figure 4D). We concluded that this result is insufficient to provide insights into the flowering mechanism in radish.

2.6. RsFT and RsSOC1-1 Floral Integrators Were Responsive to GA under Marginal Vernalization in Radish Flowering Pathways

To identify the undetected GA-specific flowering time transcripts and corroborate the putative GA-responsive flowering time DEGs, we compared transcript levels of flowering genes in the two inbred lines by performing qPCR analyses. For this purpose, we selected flowering time DEGs that exhibited different expression levels in the presence and absence of GA or were primarily involved in the flowering biological pathway and gibberellin pathways (Figure 5).

First, we identified changes in expression levels of flowering time genes involved in the GA biosynthesis and signaling pathway. The trends of GA-induced increases and decreases in gene expression levels were similar in both lines. However, the transcript fold-changes in *RsGA20ox2*, *RsKAO2*, *RsGID1A*, *RsGAI* (DELLA), and *RsGA2ox2* levels in response to GA differed between the two lines from 1.5-fold up to 5-fold. *KAO2*, *GA20ox2*, and *GA20ox3* are included in the GA biosynthesis pathway in Arabidopsis [37], whereas *GID* and *GAI* [38,39] function in GA-signaling pathway as a GA receptor and GA negative regulator, respectively, belonging to the DELLA family [40,41]. *RsBETAFRUCT4* transcript levels were similar in the presence and absence of GA, unlike a GA-specific DEG identified in RNA-seq analysis (Figure 5A).

To determine the expression levels of flowering time genes involved in the photoperiod pathway, we performed qPCR analysis of *RsELF3*, *RsLHY*, *RsGI*, *RsTEM1*, and *RsCO1 Ft* genes. *RsELF3* did not respond to GA in either line in the qPCR analysis, unlike in the RNA-seq analysis. The morning loop gene [42] *RsLHY* slightly decreased in both inbred lines when treated with GA; eventually, the differential expression levels between the two lines changed in response to GA. Similarly, qPCR analysis of *RsGI* and *RsTEM1* DEG expression levels significantly differed in the two lines in response to GA (1.5-fold up to 2.5-fold), whereas the *RsCO* transcript level did not differ between the two lines in response to GA. *GI* positively regulates *CO* and was increased by GA treatment, and their upregulation patterns were consistent with previous studies [13,43]. TEM1 directly represses expression of the GA_4 biosynthetic genes *GA3OX1* and *GA3OX2* [44], and acts to immediately repress flowering time gene expression and counteract the activator *CO* gene expression in Arabidopsis [45]. Thus, TEM1 links both photoperiod and gibberellin pathways to control flowering. *RsTEM1* was reduced in response to GA in the same manner that *GA3OX* genes were significantly upregulated in the *tem1* mutant in Arabidopsis. However, the comparative transcript levels in early- and late-flowering inbred lines did not match the bolting traits; *RsTEM1* expression was higher in NH-JS2 than in NH-JS1, and the differential expression levels were slightly larger in response to GA (Figure 5B).

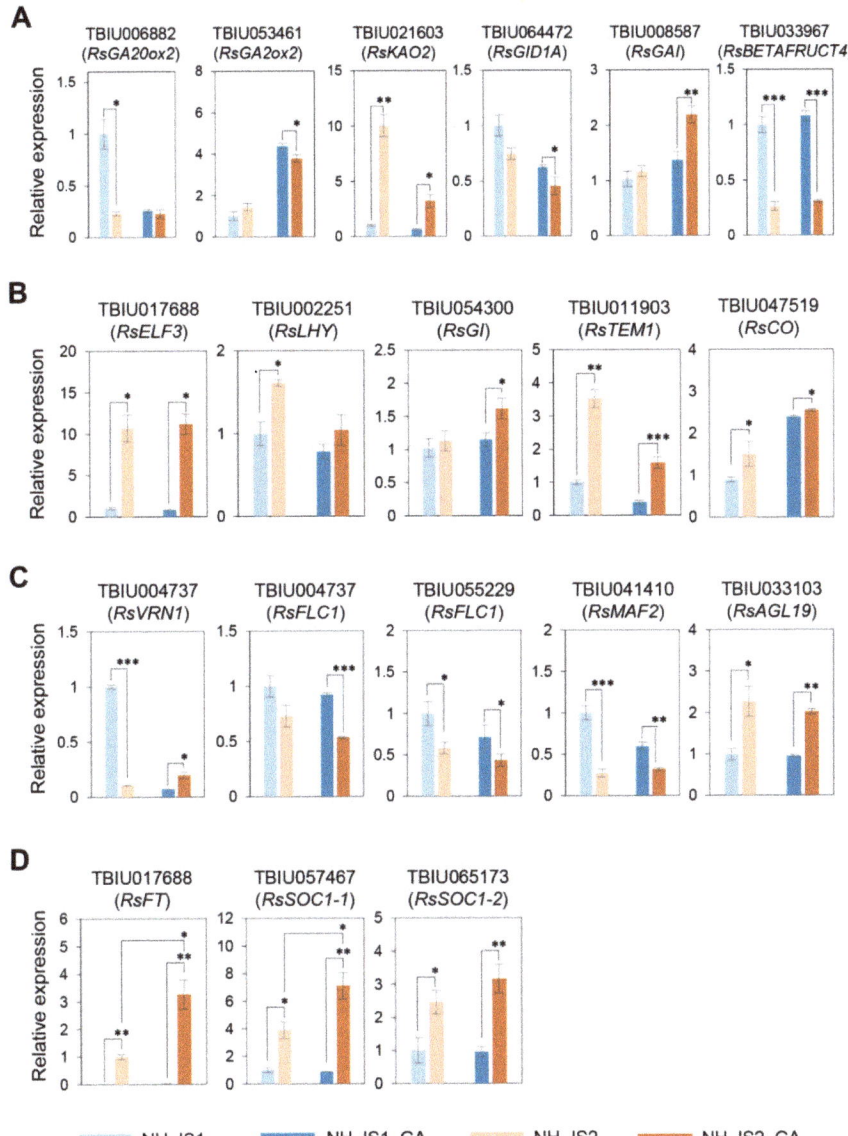

Figure 5. Quantitative PCR analysis of flowering time DEGs in the two inbred radish lines in response to gibberellin. Total RNA was isolated from shoots of NH-JS1 and NH-JS2 inbred lines 6 days after treatment with or without GA (0 or 10 mM) spray application. The complementary DNA was synthesized from total RNA. Upregulated and downregulated genes are grouped according to biological pathways determined from gene ontology analysis. Flowering time genes involved in GA pathway (**A**), photoperiod pathway (**B**), vernalization pathway (**C**) and flowering integrators (**D**). The qPCR values were normalized relative to *RsACT1* (actin) expression level. Error bars represent ± standard error of biological triplicates. Asterisks indicate statistically significant differences (NH-JS2 vs. NH-JS1 and NH-JS2_GA vs. NH-JS1_GA; Student's t test; * $p < 0.05$, ** $p < 0.01$, and *** $p < 0.005$; two-tailed Student's t-test).

Next, we quantified the expression of vernalization pathway genes. Transcript levels of the flowering repressor *RsFLC1* (TBIU004737) were lower in NH-JS2 in response to GA. Conversely, the transcript level of *RsAGL19*, a repressor of FLC1, was higher in NH-JS2 in the presence of GA. *RsFLC1* (TUBI055229) transcript levels between two lines were not significant to GA as shown in Supplementary Figure S1. *RsVRN1* expression was strongly increased in late bolting NH-JS1 plants without GA treatment, consistent with our previous RNA-seq results [32]. *RsVRN1* expression in NH-JS1 plants with GA treatment was reduced to a rarely expressed level (about 10-fold), whereas its expression was increased (approximately 2-fold) in NH-JS2 plants with GA treatment. Therefore, *RsVRN1* transcript level was higher in NH-JS2 than NH-JS1 after GA treatment, which may contribute to the observed GA hypersensitivity of NH-JS2 for early bolting and flowering. Expression of the repressor gene *RsMAF2* also responded to GA; it was less sensitive to GA stimulation in NH-JS2, whereas its transcript level was reduced about 4-fold in response to GA treatment in NH-JS1 (Figure 5C).

The expression of flowering time genes involved in the integration of multiple flowering signals was evaluated. The key floral integrator *RsFT* showed significantly different gene expression levels in response to GA in the two inbred lines. There was essentially no difference in the expression level of *RsFT* gene in NH-JS1, whereas transcript levels were more than 3-fold higher in NH-JS2 with GA than without GA. The flowering time integrator *RsSOC1-1* also showed significantly higher expression levels in response to GA treatment in NH-JS2, but were slightly reduced by GA in NH-JS1. The expression level of *RsSOC1-2*, an isoform of *SOC1*, also changed similarly as *RsSOC1-1* in response to GA between the two lines (Figure 5D). *RsSOC1-3* expression levels were not detected in this study (data not shown).

Several GA-specific flowering time genes were quantitatively identified by qPCR analysis. Some flowering time genes were not detected as flowering time- or GA-related DEGs due to RNA-seq limitations, including low abundance (*RsGA2OX2*, *RsGID1A*, *RsTEM1*, and *RsFT*) or differential expression estimation (*RsAGL19*, *RsSOC1-1*, and *RsSOC1-2*) (Supplementary Figure S5). The qPCR results revealed a transcriptional GA feedback loop that regulates GA primary response genes rather than exogenous GA treatment in radish. Expression of the essential floral genes *RsFLC1*, *RsFT*, and *RsSOC1-1*, which are generally responsible for the floral transition, was consistent with the bolting phenotype induced by GA treatment in radish.

3. Discussion

Our previous study confirmed that vernalization-mediated flowering in NH-JS2 (early-bolting phenotype) differs from that of NH-JS1 (late-bolting phenotype) [32]. Here, we investigated bolting processes closely linked to GA action under different vernalization periods. We found that NH-JS2 was more sensitive than NH-JS1 to the effect of vernalization after treatment with different concentrations of GA. To identify the GA-responsive molecular network that regulates the flowering pathway in radish, we performed RNA-seq in the two inbred lines treated with or without exogenous GA. The GA-responsive flowering time DEGs and major flowering time genes, which regulate bolting in Arabidopsis, were biologically confirmed by qPCR analysis. We suggest a gene regulatory network for controlling bolting time in response to GA in radish. Based on this model, we propose that GA promotes the vegetative-to-reproductive transition in radish by upregulating expression of the floral integrators *FT* and *SOC1* under insufficient vernalization conditions. The GA effect on flowering pathways is moderately conserved between radish and Arabidopsis.

3.1. The Two Inbred Radish Lines Display Different Bolting Times in Response to GA Treatment

GA is a positive plant growth regulator that speeds up bolting and flowering in many species [10,23,28,46] including Arabidopsis [47]. However, the effects of GA on bolting time and the molecular mechanism of GA-mediating flowering had not been reported in radish. To determine how bolting responds to GA under different vernalization periods in radish, we treated seeds of two inbred lines with GA and found that NH-JS2 displayed more bolting than NH-JS1 under insufficient vernalization conditions. In NH-JS1, bolting did not occur when vernalization was relatively short

(10 days) even with GA treatment; however, bolting did occur when vernalization was increased to 20 days with the same GA concentration (Figure 1). By contrast, the vegetative growth responses of both inbred lines to GA were essentially the same (Figure 2). GA significantly promoted bolting in NH-JS2 under insufficient vernalization conditions (10 days), but NH-JS1 displayed less bolting under the same conditions (Figure 3). This result indicates that GA has a stronger effect on the early bolting NH-JS2 line, but the late bolting NH-JS1 line reacts more sensitively to vernalization period than exogenous GA treatment. These combined results demonstrate that vernalization is an indispensable factor, whereas GA is likely to have a causal role for bolting and flowering in radish. Similarly, the grass *Lolium perenne* requires both vernalization and long-day conditions for inflorescence initiation, whereas GA promotes bolting in vernalized plants and does not affect nonvernalized plants [28]. Vernalization is essential for bolting and flowering in cauliflower (*Brassica oleracea* var. botrytis), and flowering did not occur in nonvernalized plants even with sufficient GA application [48].

3.2. GA-Responsive Flowering Time DEGs in Radish

In recent GA-responsive transcriptome studies, the most prominent gene expression changes in different tissues of *Populus tomentosa* and *Jatropha curcas* plants occurred 6 h after GA treatment [49,50]. Before performing RNA-seq analysis, we collected four samples at 6 h, and at 2, 4, and 6 days after GA treatment to check the expression levels of major flowering time genes using qPCR analysis. The differences in expression of key flowering gene such as *RsSOC1* was prominent on the sixth day after GA treatment between the two lines (Supplementary Figure S1), so RNA-seq analysis was performed on the sixth day after GA treatment (Supplementary Figure S2A). Although the bolting phenotypes induced by GA are distinct, GA-responsive flowering time DEGs were rare even though $|\log_2$ (fold-change)$| \geq 0.6$, FDR ≤ 0.01, and read count ≥ 500 were used as criteria (Figure 4). Flowering time genes could be assumed to be expressed at very low levels in response to GA; this is consistent with results in other crops as transcriptomes were analyzed without read counts involving flowering time DEGs [51–53]. We obtained 21 GA-responsive flowering time DEGs in NH-JS2 vs. NH-JS1. The numbers of upregulated vs. downregulated flowering time DEGs differed by approximately 3-fold between the two lines (upregulated vs. downregulated flowering time DEGs, 15:6) (Supplementary Table S4). Although RNA-seq is a powerful tool, there are limits for accessing low abundance transcripts, managing biological variation, and estimating differential expression. Thus, some transcripts may not be captured in the final set of reads. To overcome these limits of detection for certain genes, we conducted qPCR analysis of GA-responsive flowering time genes. Most of the detected GA-responsive flowering time DEGs were similar in the qPCR results, but fold-differences were observed due to normalization (Figure 5). In particular, qPCR analysis of the GA pathway flowering time DEGs (*RsGA20ox2*, *RsGA2ox2*, *RsKAO2*, and *RsGAI*) revealed greater differences in expression levels in response to GA in the two inbred lines than those detected in RNA-seq analysis. The qPCR results for the vernalization pathway gene *RsVRN1* showed remarkable differences in response to GA in NH-JS1, but little differences in NH-JS2. Both qPCR and RNA-seq identified significant differences in expression levels of the key floral genes *RsFT* and *RsSOC1-1* in the two lines according to the bolting traits under GA treatment (Figure 5 and Supplementary Figure S5). Our results indicate that qPCR analysis helped to provide deeper insights into the GA-responsive characteristics of radish. Further studies are needed to identify the specific functions of these genes and their molecular networks in the transition from vegetative to reproductive development in radish.

3.3. A Gene Regulatory Network Model for GA-Responsive Flowering in Radish

It is crucial to identify the molecular mechanism mediating GA responses to determine how GA application integrates the activation of flowering pathways, although it is known that GA accelerates flowering by degrading DELLA repressors. We utilized our qPCR results to develop a model for the three main flowering pathways, gibberellin, vernalization, and photoperiod, under GA treatment in radish (Figure 6). These three pathways converge on flowering integrators. By analyzing GA pathway genes, we identified that activators of GA signaling and biosynthesis were downregulated in response to GA application, whereas repressors were upregulated under low GA levels, indicating a transcriptional GA feedback mechanism functions in radish and other plant species [54]. Our results indicate that the expression of both GA biosynthesis and signaling genes was sensitive to GA, and GA maintains homeostasis through a strict mechanism in radish. In the radish photoperiod pathway, GA stimulates bolting through *RsLHY*, *RsGI*, and *RsTEM1* as activators and a repressor, respectively. LATE ELONGATED HYPOCOTYL (LHY) is a core component of the circadian oscillator [55]. GIGANTIA (GI) has roles in induction of photoperiodic flowering through FLAVIN-BINDING KELCH REPEAT, F-BOX1 (FKF1) protein interaction [56] and functions as a chaperon in the maturation of photoreceptor ZEITLUPE (ZTL) [57], whereas the TEMPRANILLO1 (TEM1) transcription factor negatively regulates GA biosynthesis and directly represses FLOWERING LOCUS T (FT) transcription [45]. Although the expression level of these genes whose expressions are regulated by circadian rhythm was determined at the one time point as 10 a.m., our result shows that GA stimulates the photoperiod pathway activator *RsGI* and represses the *RsTEM1* floral repressor in radish. In the radish vernalization pathway, *RsVRN1* was the most GA-responsive and unique gene that displayed expression pattern changes under GA treatment, consistent with the bolting characters in both lines. The transcript levels declined in response to GA in the late-bolting NH JS1 line (approximately 14-fold change), whereas the early-bolting line NH-JS2 showed inversely increased expression in response to GA (approximately 2-fold change). Vernalization1 (VRN1) is central to the vernalization response and important to maintain repression of the flowering repressor VRN2, ultimately promoting the upregulation of flowering time expression in Arabidopsis [58]. This result suggests that changes in *RsVRN1* expression in response to GA could largely induce the transition from vegetative to reproductive growth in radish under marginal vernalization conditions. Each flowering pathway was distinct or interrelated, and eventually converged to the floral integrators FT, SOC1, and LFY. Ultimately, upregulation of *RsFT* and *RsSOC1-1* in response to GA is sufficient to induce bolting. *RsLFY* was rarely expressed in both RNA-seq and qPCR analyses in this study. Our RNA-seq results confirmed that most of the flowering gene expression levels were very low, which could be due to two reasons: (1) We performed RNA-seq analysis of young radish shoots, and (2) GA-mediated flowering induction is likely to be effective even at very low levels of gene expression. This study identified transcripts that have differentially low expression levels in response to GA in early-bolting and late-bolting inbred radish lines. These DEGs may reveal crucial information about GA-mediated flowering in radish (Figure 6).

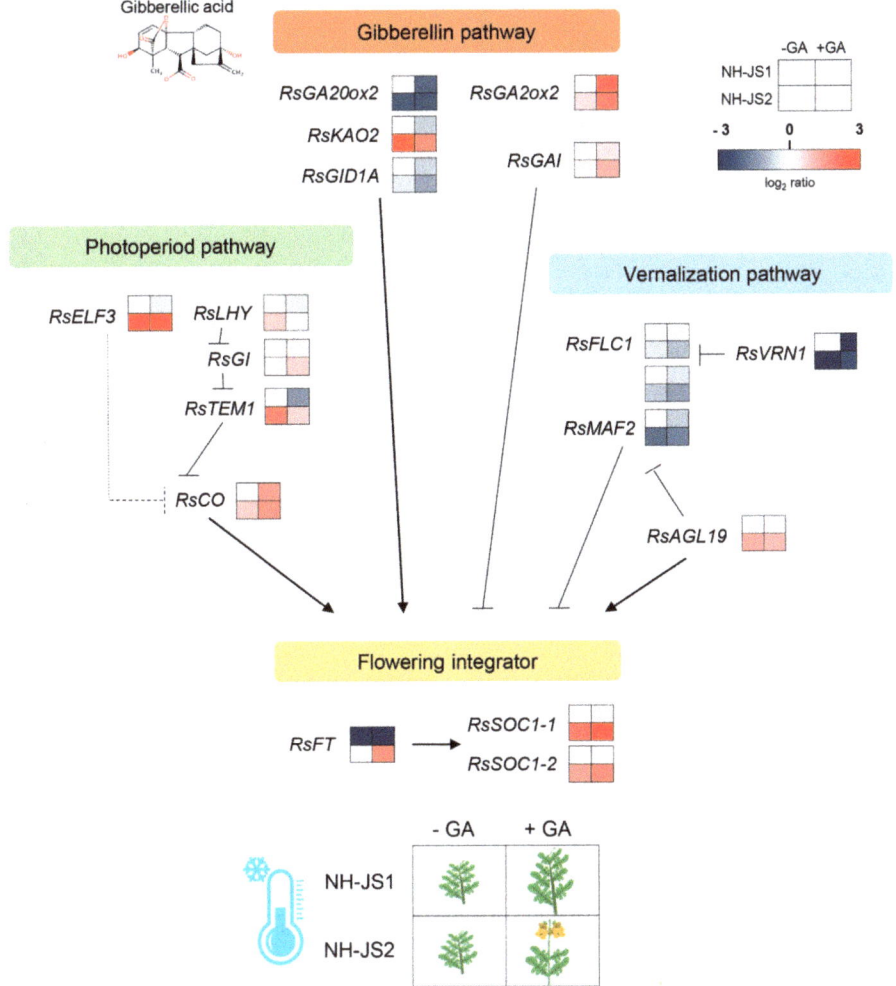

Figure 6. Gene regulatory network controlling gibberellin-accelerated flowering in radish. The illustration maps the regulatory network of flowering time genes in NH-JS1 (late-bolting) and NH-JS2 (early-bolting) plants treated with GA. The network is based on data confirmed by qPCR. Gene expression levels were normalized relative to the expression levels in non-GA-treated NH-JS1 plants (for *RsFT*, non-GA-treated NH-JS2 plant data analysis included log$_2$ ratio.). Red indicates higher expression levels and blue indicates lower expression levels relative to non-GA-treated NH-JS1 (*RsFT* expression is relative to non-GA-treated NH-JS2). Arrows indicate transcriptional activation; bars indicate transcriptional repression.

4. Materials and Methods

4.1. Plant Materials, Exogenous GA Treatment, and Bolting Trait Analysis

The NH-JS1 (late-bolting) and NH-JS2 (early-bolting) radish inbred lines were developed by NongHyup Seed (Geonggi-do, Anseong, Korea) [32]. Seeds of each line were sterilized by soaking in 70% ethanol and 10% chlorine bleach for 30 min to disinfect the seed coat, and then washed with sterile distilled water at least three times. To investigate the effect of exogenous GA on seedlings, seeds of each line were sown on filter paper treated with 5 mL of 0, 0.1, or 10 mM GA (Duchefa Biochemie, Haarlem, The Netherlands) at 25 °C for 1 day. Seedlings with a length of 1–2 mm were grown in the growth room under long-day conditions (25 °C, 16 h light/8 h dark photoperiod at 100 µmol m^{-2}s^{-1}) for 14 days. Thereafter, the seedlings were vernalized in a cold room (5 ± 1 °C, 12 h light/12 h dark at 100 µmol m^{-2}s^{-1}) for 0, 10, and 20 days, and then transferred to sterilized soil and grown under normal growth conditions. Twenty seedlings were used for each bolting test. To test the effect of exogenous GA on young plants, 4-day-old seedlings were vernalized (as described above) for 10 days and then transferred to the growth room for 2 weeks to acclimatize. After acclimation, 10 mM GA solution was sprayed directly on the leaves, and plants were grown in the growth room for up to 50 days after spraying. Plants were examined for bolting at 0, 15, 20, 25, 30, 35, and 40 days after vernalization (DAV); more than 20 plants were used for each test. The percentage of bolted plants was calculated by counting the number of plants with floral axis lengths ≥1 cm relative to the plants without floral axes. The bolting percentages for each inbred line after GA treatment of seeds were calculated at 25, 30, and 35 days after each vernalization treatment of 0, 10, and 20 days. The bolting percentages for each inbred line after GA treatment of young plant leaves were calculated at 0, 15, 20, 25, 30, 35, and 40 days after the 10-day vernalization treatment.

4.2. Evaluating the GA Effect on Hypocotyl Elongation

Twenty seeds of each inbred line were disinfected, sown in sterilized soil, placed in the dark in a cold room at 5 ± 1 °C for 3 days to enhance germination, and then transferred to a growth room. Radish sprouts with hypocotyls longer than 0.5 cm in length were transferred to another pot containing sterilized soil in the growth room. Then, 1 mL of 0, 1, or 10 mM GA solution was directly poured over each sprout. Hypocotyl lengths were measured every day for 6 days after the GA treatment. More than 20 seedlings of each inbred line were used for hypocotyl measurement at each GA concentration.

4.3. Preparation and Sequencing of the RNA-Seq Library

A total of eight shoot tissue samples (two inbred lines × two treatments × two biological replicates) were collected at the same point in the light/dark cycle as 10 a.m. time point, immediately frozen in liquid nitrogen, and stored at −80 °C until further processing. Shoot tissues from three different plants were pooled to obtain sufficient RNA for each extraction. For RNA-seq library construction, total RNA was isolated from shoot tissue and cDNA was synthesized as described by Jung et al. [32]. The RNA-seq libraries were created using the Illumina TruSeq RNA Library Prep Kit (Illumina, San Diego, CA, USA) according to the manufacturer's protocol. The RNA-seq library was PCR-amplified and sequenced on an Illumina HiSeq 2000 platform. A 101-bp paired-end sequencing protocol was used, and two biological repeats were performed for each sample. All raw, read data generated in this study were deposited in the Gene Expression Omnibus (GEO) functional genomics data repository of National Center for Biotechnology Information under accession number GSE125875: lists the GEO DataSeries.

4.4. Reference-Guided Assembly and Mapping of the Radish Transcriptome

Raw sequencing data were filtered to standard Illumina pipeline RNA-seq parameters. Paired-end reads were quality trimmed, and adapter contamination, low-quality parts, and N-base reads were removed. Reads that fell below a Phred quality score ($Q \leq 20$) and reads shorter than 25 base pairs (bp) were discarded. These steps were performed using the DynamicTrim and LengthSort programs of the SolexaQA (v.1.13) package [59]. Next, the purified paired-end reads were pooled and mapped against available radish reference gene datasets. A reference-based assembly was performed by utilizing 46,512 genes from the coding sequence regions of the radish reference genome to obtain an assembly dataset [30]. Bowtie2 (v.2.1.0) was used to map the purified datasets [60]. The program allows only remarkable mapping, which has a maximum of two mismatches. Otherwise, the default options were used. The expression levels of each sample were determined using in-house scripts. Read counts of each gene were normalized with respect to library size and counted to the nearest whole number.

4.5. Functional Annotation Analysis

RNA-seq transcripts were annotated by comparison with gene sequences in the Phytozome database using Basic Local Alignment Search Tool: protein BLAST (BLASTp) with expect values (E-values) that were at least higher than $1E^{-10}$ (BLAST v.2.2.28+) [61]. The Gene Ontology (GO) database was utilized for GO analysis, and the transcripts were annotated through the GO database using BLASTP (E-value $\leq 1E^{-06}$). GO term annotation was conducted using GO classification results from the Map2Slim.pl script. Protein sequences were annotated with the highest sequence similarities and cutoffs, and retrieved for analysis. Data for Annotation, Visulization and Integrated Discovery (DAVID) was used for functional enrichment analysis [62]. The transcript lists also were annotated using the The Arabidopsis Information Resource database and clarified according to default criteria (counts ≥ 2 and the Expression Analysis Systematic Explorer score ≤ 0.1). The Kyoto Encyclopedia of Gene and Genome (KEGG) pathways database was used to analyze the sequences using the single-directional best-hit method and the KEGG Automatic Annotation Server [63]. KEGG enrichment analysis was performed as described previously [55].

4.6. Analysis of DEGs

Gene expression data were generated from eight samples of the two inbred lines. To identify DEGs, the plant samples were treated with GA for 6 days with 10 days of vernalization. Raw counts were normalized and analyzed using the differntial gene expression analysis based on the negative binomial distribution (DESeq) library in R (v3.2) [64]. Then, DEGs were considered to display fold-change more than $|\log_2 \text{(fold-change)}| \geq 0.6$ and were filtered by requiring the *p*-value adjustment to be ≤ 0.01. The control was NH-JS1 (without GA).

4.7. Identification of Flowering Time- and GA-Related Genes in Radish

To identify flowering time- and GA pathway-related genes in our transcriptomes, two sets of 218 flowering time and 125 GA pathway-related genes were selected as reference sets based on published literature and studies in *Arabidopsis thaliana* [65–68] as described previously [32]. Published sequences were obtained from the TAIR database based on Arabidopsis accession numbers for flowering time- and GA-related genes. BLASTn was used to query the 218 flowering time- and 125 GA-related genes against the assembled 71,188 genes of radish. Top hits were filtered based on the highest percentage of hit coverage and sequence similarity. Cutoffs were E-values $\leq 1E^{-25}$ and identity $\geq 65\%$. The flowering time- and GA-related genes in Arabidopsis were compared with flowering time- and GA-related gene sequences in radish using BLASTn (E-values $\leq 1E^{-25}$, identity $\geq 70\%$).

4.8. Quantitative PCR (qPCR) Analysis

To validate the DEGs identified in RNA-seq analysis, we conducted qPCR analysis. Total RNA was isolated from radish with or without GA treatment using RNAiso Plus (TaKaRa, Tokyo, Japan). Total RNA with RNase-free DNase I (Fermentas, Burlington, Canada) was used for cDNA synthesis (RevertAid First-Strand cDNA Synthesis Kit, Fermentas). Then, qPCR was performed in a CFX Connect™ Real-Time PCR System (Bio-Rad, Hercules, CA, USA) using SYBR Premix Ex-Taq (TaKaRa, Tokyo, Japan) according to the manufacturer's instructions. Relative expression levels were obtained after normalization with radish actin gene (*RsACT*) expression levels. All qPCR experiments were performed using the *flowering time* gene-specific primer set (Supplementary Table S6) with two biological replicates, each with three technical repeats, under the same conditions.

Supplementary Materials: The following are available online at http://www.mdpi.com/2223-7747/9/5/594/s1: Figure S1: The qPCR analysis of flowering time DEGs in response to GA between the two inbred lines. Figure S2: Experimental design for RNA-seq analysis. Figure S3: DEGs' analysis by gibberellin treatment on each line. Figure S4: Comparative Kyoto Encyclopedia Genes and Genomes pathways of DEGs in two inbred lines under gibberellin treatment. Figure S5: RNA-seq validation of flowering time genes. Table S1: Cleaned short reads statistics after trimming of RNA sequencing data. Table S2: Statistics of reads mapping to Theragen Bio Institute (TBI)—unigenes. Table S3: Radish homologs of GA-specific genes. Table S4: Flowering time-specific DEGs on GA treatment. Table S5: Gibberellin-specific DEGs on GA treatment. Table S6: Gene-specific primer information for qPCR analysis.

Author Contributions: H.S.C. and Y.-S.K. conceived and designed the study and wrote the manuscript. H.J. performed bioinformatic data analysis and wrote the manuscript. S.H.J. performed gibberellin-responsive phenotyping and wrote the manuscript. W.Y.J. analyzed bioinformatics data. H.J.P., S.Y.S., and A.L. conducted gene expression analysis. J.S.M. and J.-K.K. advised data analysis. All authors read and approved the manuscript.

Funding: This research was funded by the Next Generation New Plant Breed Technology Program (no. PJ014809) of the Rural Development Administration and by the Korea Research Institute of Bioscience and Biotechnology Research Initiative Program.

Acknowledgments: The authors acknowledge breeders of NongHyup Seeds Co., for providing inbred lines.

Conflicts of Interest: The authors declare no conflicts of interest.

References

1. Kasahara, H.; Hanada, A.; Kuzuyama, T.; Takagi, M.; Kamiya, Y.; Yamaguchi, S. Contribution of the mevalonate and methylerythritol phosphate pathways to the biosynthesis of gibberellins in arabidopsis. *J. Biol. Chem* **2002**, *277*, 45188–45194. [CrossRef] [PubMed]
2. Goldberg-Moeller, R.; Shalom, L.; Shlizerman, L.; Samuels, S.; Zur, N.; Ophir, R.; Blumwald, E.; Sadka, A. Effects of gibberellin treatment during flowering induction period on global gene expression and the transcription of flowering-control genes in citrus buds. *Plant Sci. An. Int. J. Exp. Plant Biol.* **2013**, *198*, 46–57. [CrossRef] [PubMed]
3. Sun, T.P.; Gubler, F. Molecular mechanism of gibberellin signaling in plants. *Annu. Rev. Plant Biol.* **2004**, *55*, 197–223. [CrossRef] [PubMed]
4. Daviere, J.M.; Achard, P. A pivotal role of dellas in regulating multiple hormone signals. *Mol. Plant* **2016**, *9*, 10–20. [CrossRef] [PubMed]
5. Xu, Q.; Krishnan, S.; Merewitz, E.; Xu, J.; Huang, B. Gibberellin-regulation and genetic variations in leaf elongation for tall fescue in association with differential gene expression controlling cell expansion. *Sci. Rep.* **2016**, *6*, 30258. [CrossRef]
6. Achard, P.; Gusti, A.; Cheminant, S.; Alioua, M.; Dhondt, S.; Coppens, F.; Beemster, G.T.; Genschik, P. Gibberellin signaling controls cell proliferation rate in arabidopsis. *Curr. Biol.* **2009**, *19*, 1188–1193. [CrossRef]
7. Achard, P.; Cheng, H.; De Grauwe, L.; Decat, J.; Schoutteten, H.; Moritz, T.; Van Der Straeten, D.; Peng, J.; Harberd, N.P. Integration of plant responses to environmentally activated phytohormonal signals. *Science (N. Y.)* **2006**, *311*, 91–94. [CrossRef]
8. Achard, P.; Gong, F.; Cheminant, S.; Alioua, M.; Hedden, P.; Genschik, P. The cold-inducible cbf1 factor-dependent signaling pathway modulates the accumulation of the growth-repressing della proteins via its effect on gibberellin metabolism. *Plant. Cell* **2008**, *20*, 2117–2129. [CrossRef]
9. Magome, H.; Yamaguchi, S.; Hanada, A.; Kamiya, Y.; Oda, K. The ddf1 transcriptional activator upregulates expression of a gibberellin-deactivating gene, ga2ox7, under high-salinity stress in arabidopsis. *Plant J.* **2008**, *56*, 613–626. [CrossRef]
10. Hedden, P.; Thomas, S.G. Gibberellin biosynthesis and its regulation. *Biochem. J.* **2012**, *444*, 11–25. [CrossRef]
11. Thomas, S.G.; Rieu, I.; Steber, C.M. Gibberellin metabolism and signaling. *Vitam. Horm.* **2005**, *72*, 289–338. [PubMed]
12. Murase, K.; Hirano, Y.; Sun, T.P.; Hakoshima, T. Gibberellin-induced della recognition by the gibberellin receptor gid1. *Nature* **2008**, *456*, 459–463. [CrossRef]
13. Shimada, A.; Ueguchi-Tanaka, M.; Nakatsu, T.; Nakajima, M.; Naoe, Y.; Ohmiya, H.; Kato, H.; Matsuoka, M. Structural basis for gibberellin recognition by its receptor gid1. *Nature* **2008**, *456*, 520–523. [CrossRef] [PubMed]

14. Itoh, H.; Matsuoka, M.; Steber, C.M. A role for the ubiquitin-26s-proteasome pathway in gibberellin signaling. *Trends Plant. Sci.* **2003**, *8*, 492–497. [CrossRef]
15. Du, R.; Niu, S.; Liu, Y.; Sun, X.; Porth, I.; El-Kassaby, Y.A.; Li, W. The gibberellin gid1-della signalling module exists in evolutionarily ancient conifers. *Sci. Rep.* **2017**, *7*, 16637. [CrossRef] [PubMed]
16. Hedden, P. The genes of the green revolution. *Trends Genet.* **2003**, *19*, 5–9. [CrossRef]
17. Peng, J.; Richards, D.E.; Hartley, N.M.; Murphy, G.P.; Devos, K.M.; Flintham, J.E.; Beales, J.; Fish, L.J.; Worland, A.J.; Pelica, F.; et al. 'Green revolution' genes encode mutant gibberellin response modulators. *Nature* **1999**, *400*, 256–261. [CrossRef]
18. Elias, A.A.; Busov, V.B.; Kosola, K.R.; Ma, C.; Etherington, E.; Shevchenko, O.; Gandhi, H.; Pearce, D.W.; Rood, S.B.; Strauss, S.H. Green revolution trees: Semidwarfism transgenes modify gibberellins, promote root growth, enhance morphological diversity, and reduce competitiveness in hybrid poplar. *Plant Physiol.* **2012**, *160*, 1130–1144. [CrossRef]
19. Zhao, M.; Liu, B.; Wu, K.; Ye, Y.; Huang, S.; Wang, S.; Wang, Y.; Han, R.; Liu, Q.; Fu, X.; et al. Regulation of osmir156h through alternative polyadenylation improves grain yield in rice. *PLoS ONE* **2015**, *10*, e0126154. [CrossRef]
20. Komeda, Y. Genetic regulation of time to flower in arabidopsis thaliana. *Annu. Rev. Plant. Biol.* **2004**, *55*, 521–535. [CrossRef]
21. Davis, S.J. Integrating hormones into the floral-transition pathway of arabidopsis thaliana. *Plant. Cell Environ.* **2009**, *32*, 1201–1210. [CrossRef] [PubMed]
22. Simpson, G.G.; Dean, C. Arabidopsis, the rosetta stone of flowering time? *Science (N. Y.)* **2002**, *296*, 285–289. [CrossRef] [PubMed]
23. Mutasa-Gottgens, E.; Hedden, P. Gibberellin as a factor in floral regulatory networks. *J. Exp. Bot.* **2009**, *60*, 1979–1989. [CrossRef] [PubMed]
24. Wilkie, J.D.; Sedgley, M.; Olesen, T. Regulation of floral initiation in horticultural trees. *J. Exp. Bot.* **2008**, *59*, 3215–3228. [CrossRef]
25. Bangerth, K.F. Floral induction in mature, perennial angiosperm fruit trees: Similarities and discrepancies with annual/biennial plants and the involvement of plant hormones. *Sci. Hort.* **2009**, *122*, 153–163. [CrossRef]
26. Achard, P.; Herr, A.; Baulcombe, D.C.; Harberd, N.P. Modulation of floral development by a gibberellin-regulated microrna. *Development* **2004**, *131*, 3357–3365. [CrossRef]
27. Moon, J.; Suh, S.S.; Lee, H.; Choi, K.R.; Hong, C.B.; Paek, N.C.; Kim, S.G.; Lee, I. The soc1 mads-box gene integrates vernalization and gibberellin signals for flowering in arabidopsis. *Plant. J.* **2003**, *35*, 613–623. [CrossRef]
28. Macmillan, C.P.; Blundell, C.A.; King, R.W. Flowering of the grass lolium perenne: Effects of vernalization and long days on gibberellin biosynthesis and signaling. *Plant. Physiol.* **2005**, *138*, 1794–1806. [CrossRef]
29. Kitashiba, H.; Li, F.; Hirakawa, H.; Kawanabe, T.; Zou, Z.; Hasegawa, Y.; Tonosaki, K.; Shirasawa, S.; Fukushima, A.; Yokoi, S.; et al. Draft sequences of the radish (raphanus sativus l.) genome. *DNA Res.* **2014**, *21*, 481–490. [CrossRef]
30. Mun, J.H.; Chung, H.; Chung, W.H.; Oh, M.; Jeong, Y.M.; Kim, N.; Ahn, B.O.; Park, B.S.; Park, S.; Lim, K.B.; et al. Construction of a reference genetic map of raphanus sativus based on genotyping by whole-genome resequencing. *Theor. Appl. Genet.* **2015**, *128*, 259–272. [CrossRef]
31. Nie, S.; Li, C.; Xu, L.; Wang, Y.; Huang, D.; Muleke, E.M.; Sun, X.; Xie, Y.; Liu, L. De novo transcriptome analysis in radish (raphanus sativus l.) and identification of critical genes involved in bolting and flowering. *BMC Genom.* **2016**, *17*, 389. [CrossRef] [PubMed]
32. Jung, W.Y.; Park, H.J.; Lee, A.; Lee, S.S.; Kim, Y.S.; Cho, H.S. Identification of flowering-related genes responsible for differences in bolting time between two radish inbred lines. *Front. Plant Sci.* **2016**, *7*, 1844. [CrossRef] [PubMed]
33. Yi, G.; Park, H.; Kim, J.-S.; Chae, W.; Park, S.; Huh, J. Identification of three flowering locus c genes responsible for vernalization response in radish (raphanus sativus l.). *Hortic. Environ. Biotechnol.* **2014**, *55*, 548–556. [CrossRef]
34. Erwin, J.E.; Warner, R.M.; Smith, A.G. Vernalization, photoperiod and ga3 interact to affect flowering of japanese radish (raphanus sativus chinese radish jumbo scarlet). *Physiol. Plant.* **2002**, *115*, 298–302. [CrossRef] [PubMed]

35. Han, Y.; Wan, H.; Cheng, T.; Wang, J.; Yang, W.; Pan, H.; Zhang, Q. Comparative rna-seq analysis of transcriptome dynamics during petal development in rosa chinensis. *Sci. Rep.* **2017**, *7*, 43382. [CrossRef] [PubMed]
36. Qu, J.; Kang, S.G.; Hah, C.; Jang, J.C. Molecular and cellular characterization of ga-stimulated transcripts gasa4 and gasa6 in arabidopsis thaliana. *Plant Sci. An. Int. J. Exp. Plant Biol.* **2016**, *246*, 1–10. [CrossRef] [PubMed]
37. Toh, S.; Imamura, A.; Watanabe, A.; Nakabayashi, K.; Okamoto, M.; Jikumaru, Y.; Hanada, A.; Aso, Y.; Ishiyama, K.; Tamura, N.; et al. High temperature-induced abscisic acid biosynthesis and its role in the inhibition of gibberellin action in arabidopsis seeds. *Plant. Physiol.* **2008**, *146*, 1368–1385. [CrossRef]
38. Peng, J.; Harberd, N.P. Gibberellin deficiency and response mutations suppress the stem elongation phenotype of phytochrome-deficient mutants of arabidopsis. *Plant. Physiol.* **1997**, *113*, 1051–1058. [CrossRef]
39. Peng, J.; Harberd, N.P. Derivative alleles of the arabidopsis gibberellin-insensitive (gai) mutation confer a wild-type phenotype. *The Plant. Cell* **1993**, *5*, 351–360. [CrossRef]
40. Silverstone, A.L.; Tseng, T.S.; Swain, S.M.; Dill, A.; Jeong, S.Y.; Olszewski, N.E.; Sun, T.P. Functional analysis of spindly in gibberellin signaling in arabidopsis. *Plant. Physiol.* **2007**, *143*, 987–1000. [CrossRef]
41. Sun, T.P. Gibberellin-gid1-della: A pivotal regulatory module for plant growth and development. *Plant. Physiol.* **2010**, *154*, 567–570. [CrossRef] [PubMed]
42. Takata, N.; Saito, S.; Saito, C.T.; Nanjo, T.; Shinohara, K.; Uemura, M. Molecular phylogeny and expression of poplar circadian clock genes, lhy1 and lhy2. *New Phytol.* **2009**, *181*, 808–819. [CrossRef] [PubMed]
43. Mouradov, A.; Cremer, F.; Coupland, G. Control of flowering time: Interacting pathways as a basis for diversity. *Plant. Cell* **2002**, *14* (Suppl. 1), S111–S130. [CrossRef] [PubMed]
44. Osnato, M.; Castillejo, C.; Matias-Hernandez, L.; Pelaz, S. Tempranillo genes link photoperiod and gibberellin pathways to control flowering in arabidopsis. *Nat. Commun.* **2012**, *3*, 808. [CrossRef] [PubMed]
45. Castillejo, C.; Pelaz, S. The balance between constans and tempranillo activities determines ft expression to trigger flowering. *Curr. Biol. CB* **2008**, *18*, 1338–1343. [CrossRef]
46. King, R.W.; Moritz, T.; Evans, L.T.; Junttila, O.; Herlt, A.J. Long-day induction of flowering in lolium temulentum involves sequential increases in specific gibberellins at the shoot apex. *Plant. Physiol.* **2001**, *127*, 624–632. [CrossRef]
47. Eriksson, S.; Bohlenius, H.; Moritz, T.; Nilsson, O. Ga4 is the active gibberellin in the regulation of leafy transcription and arabidopsis floral initiation. *Plant. Cell* **2006**, *18*, 2172–2181. [CrossRef]
48. Guo, D.-P.; Ali Shah, G.; Zeng, G.-W.; Zheng, S.-J. The interaction of plant growth regulators and vernalization on the growth and flowering of cauliflower (brassica oleracea var. Botrytis). *Plant. Growth Regul.* **2004**, *43*, 163–171. [CrossRef]
49. Xie, J.; Tian, J.; Du, Q.; Chen, J.; Li, Y.; Yang, X.; Li, B.; Zhang, D. Association genetics and transcriptome analysis reveal a gibberellin-responsive pathway involved in regulating photosynthesis. *J. Exp. Bot.* **2016**, *67*, 3325–3338. [CrossRef]
50. Hui, W.K.; Wang, Y.; Chen, X.Y.; Zayed, M.Z.; Wu, G.J. Analysis of transcriptional responses of the inflorescence meristems in jatropha curcas following gibberellin treatment. *Int. J. Mol. Sci.* **2018**, *19*, 432. [CrossRef]
51. Singh, V.K.; Jain, M. Transcriptome profiling for discovery of genes involved in shoot apical meristem and flower development. *Genom Data* **2014**, *2*, 135–138. [CrossRef] [PubMed]
52. Zhang, C.; Lin, C.; Fu, F.; Zhong, X.; Peng, B.; Yan, H.; Zhang, J.; Zhang, W.; Wang, P.; Ding, X.; et al. Comparative transcriptome analysis of flower heterosis in two soybean f1 hybrids by rna-seq. *PLoS ONE* **2017**, *12*, e0181061. [CrossRef]
53. Liu, K.; Feng, S.; Pan, Y.; Zhong, J.; Chen, Y.; Yuan, C.; Li, H. Transcriptome analysis and identification of genes associated with floral transition and flower development in sugar apple (annona squamosa l.). *Front. Plant Sci.* **2016**, *7*, 1695. [CrossRef] [PubMed]
54. Olszewski, N.; Sun, T.P.; Gubler, F. Gibberellin signaling: Biosynthesis, catabolism, and response pathways. *Plant. Cell* **2002**, *14* (Suppl. 1), S61–S80. [CrossRef]
55. Alabadi, D.; Oyama, T.; Yanovsky, M.J.; Harmon, F.G.; Mas, P.; Kay, S.A. Reciprocal regulation between toc1 and lhy/cca1 within the arabidopsis circadian clock. *Science (N. Y.)* **2001**, *293*, 880–883. [CrossRef]
56. Sawa, M.; Nusinow, D.A.; Kay, S.A.; Imaizumi, T. Fkf1 and gigantea complex formation is required for day-length measurement in arabidopsis. *Science* **2007**, *318*, 261–265. [CrossRef]

57. Cha, J.Y.; Kim, J.; Kim, T.S.; Zeng, Q.; Wang, L.; Lee, S.Y.; Kim, W.Y.; Somers, D.E. Gigantea is a co-chaperone which facilitates maturation of zeitlupe in the arabidopsis circadian clock. *Nat. Commun.* **2017**, *8*, 3. [CrossRef] [PubMed]
58. Levy, Y.Y.; Mesnage, S.; Mylne, J.S.; Gendall, A.R.; Dean, C. Multiple roles of arabidopsis vrn1 in vernalization and flowering time control. *Science* **2002**, *297*, 243–246. [CrossRef] [PubMed]
59. Cox, M.P.; Peterson, D.A.; Biggs, P.J. Solexaqa: At-a-glance quality assessment of illumina second-generation sequencing data. *BMC Bioinform.* **2010**, *11*, 485. [CrossRef] [PubMed]
60. Langmead, B.; Salzberg, S.L. Fast gapped-read alignment with bowtie 2. *Nat. Methods* **2012**, *9*, 357–359. [CrossRef] [PubMed]
61. Altschul, S.F.; Madden, T.L.; Schaffer, A.A.; Zhang, J.; Zhang, Z.; Miller, W.; Lipman, D.J. Gapped blast and psi-blast: A new generation of protein database search programs. *Nucleic Acids Res.* **1997**, *25*, 3389–3402. [CrossRef] [PubMed]
62. Huang da, W.; Sherman, B.T.; Lempicki, R.A. Systematic and integrative analysis of large gene lists using david bioinformatics resources. *Nat. Protoc.* **2009**, *4*, 44–57. [CrossRef] [PubMed]
63. Moriya, Y.; Itoh, M.; Okuda, S.; Yoshizawa, A.C.; Kanehisa, M. Kaas: An automatic genome annotation and pathway reconstruction server. *Nucleic Acids Res.* **2007**, *35*, W182–W185. [CrossRef] [PubMed]
64. Anders, S.; Huber, W. Differential expression analysis for sequence count data. *Genome Biol.* **2010**, *11*, R106. [CrossRef]
65. Nilsson, O.; Lee, I.; Blazquez, M.A.; Weigel, D. Flowering-time genes modulate the response to leafy activity. *Genetics* **1998**, *150*, 403–410.
66. Amasino, R.M.; Michaels, S.D. The timing of flowering. *Plant. Physiol.* **2010**, *154*, 516–520. [CrossRef]
67. Yamauchi, Y.; Ogawa, M.; Kuwahara, A.; Hanada, A.; Kamiya, Y.; Yamaguchi, S. Activation of gibberellin biosynthesis and response pathways by low temperature during imbibition of arabidopsis thaliana seeds. *Plant. Cell* **2004**, *16*, 367–378. [CrossRef]
68. Colebrook, E.H.; Thomas, S.G.; Phillips, A.L.; Hedden, P. The role of gibberellin signalling in plant responses to abiotic stress. *J. Exp. Biol.* **2014**, *217*, 67–75. [CrossRef]

© 2020 by the authors. Licensee MDPI, Basel, Switzerland. This article is an open access article distributed under the terms and conditions of the Creative Commons Attribution (CC BY) license (http://creativecommons.org/licenses/by/4.0/).

Article

Genetic Diversity of Genes Controlling Unilateral Incompatibility in Japanese Cultivars of Chinese Cabbage

Yoshinobu Takada [1,*], Atsuki Mihara [2], Yuhui He [2], Haolin Xie [2], Yusuke Ozaki [2], Hikari Nishida [2], Seongmin Hong [3], Yong-Pyo Lim [3], Seiji Takayama [4], Go Suzuki [2,*] and Masao Watanabe [1,*]

1 Graduate School of Life Sciences, Tohoku University, Sendai 980-8577, Japan
2 Division of Natural Science, Osaka Kyoiku University, Kashiwara 582-8582, Japan; am.osaka.kyoiku@gmail.com (A.M.); ukiko0620@gmail.com (Y.H.); xiehaolin20@gmail.com (H.X.); yu.oz.7991@gmail.com (Y.O.); hkr.disney.12121998@gmail.com (H.N.)
3 Department of Horticulture, College of Agriculture and Life Science, Chungnam National University, Daejeon 34134, Korea; sungmin201@cnu.ac.kr (S.H.); yplim@cnu.ac.kr (Y.-P.L.)
4 Department of Applied Biological Chemistry, Graduate School of Agricultural and Life Sciences, The University of Tokyo, Tokyo 113-8657, Japan; a-taka@mail.ecc.u-tokyo.ac.jp
* Correspondence: ytakada@ige.tohoku.ac.jp (Y.T.); gsuzuki@cc.osaka-kyoiku.ac.jp (G.S.); nabe@ige.tohoku.ac.jp (M.W.)

Citation: Takada, Y.; Mihara, A.; He, Y.; Xie, H.; Ozaki, Y.; Nishida, H.; Hong, S.; Lim, Y.-P.; Takayama, S.; Suzuki, G.; et al. Genetic Diversity of Genes Controlling Unilateral Incompatibility in Japanese Cultivars of Chinese Cabbage. *Plants* 2021, *10*, 2467. https://doi.org/10.3390/plants10112467

Academic Editor: Javier Rodrigo

Received: 30 September 2021
Accepted: 12 November 2021
Published: 15 November 2021

Publisher's Note: MDPI stays neutral with regard to jurisdictional claims in published maps and institutional affiliations.

Copyright: © 2021 by the authors. Licensee MDPI, Basel, Switzerland. This article is an open access article distributed under the terms and conditions of the Creative Commons Attribution (CC BY) license (https://creativecommons.org/licenses/by/4.0/).

Abstract: In recent years, unilateral incompatibility (UI), which is an incompatibility system for recognizing and rejecting foreign pollen that operates in one direction, has been shown to be closely related to self-incompatibility (SI) in *Brassica rapa*. The stigma- and pollen-side recognition factors (*SUI1* and *PUI1*, respectively) of this UI are similar to those of SI (stigma-side *SRK* and pollen-side *SP11*), indicating that *SUI1* and *PUI1* interact with each other and cause pollen-pistil incompatibility only when a specific genotype is pollinated. To clarify the genetic diversity of *SUI1* and *PUI1* in Japanese *B. rapa*, here we investigated the UI phenotype and the *SUI1/PUI1* sequences in Japanese commercial varieties of Chinese cabbage. The present study showed that multiple copies of nonfunctional *PUI1* were located within and in the vicinity of the *UI* locus region, and that the functional *SUI1* was highly conserved in Chinese cabbage. In addition, we found a novel nonfunctional *SUI1* allele with a dominant negative effect on the functional *SUI1* allele in the heterozygote.

Keywords: allelic diversity; *Brassica rapa*; Chinese cabbage; dominant negative effect; gene duplication; pollen-stigma interaction; self-incompatibility; unilateral incompatibility

1. Introduction

Most Japanese cultivars of Chinese cabbage (*Brassica rapa* L.) are F$_1$ hybrids. Traditionally, their seeds have been produced using the *Brassica* self-incompatibility (SI) system. The SI system in *Brassica* is sporophytically controlled by a single *S*-locus with highly variable, multiple alleles [1]. The *S*-locus region contains two genes, *SRK* and *SP11/SCR*, which correspond to female and male *S* determinants, respectively [2]. *SRK* encodes a transmembrane receptor kinase, which is expressed specifically in stigma, and *SP11/SCR* encodes a small cysteine-rich ligand for SRK, which is localized on the pollen coat [3–5]. The *S*-haplotype-specific interaction of SP11 and the extracellular domain of SRK induces the SI reaction, in which the self-pollen fails to germinate or penetrate into the stigma [6]. The number of *S*-haplotypes has been estimated to be more than 100 in *B. rapa* [7–9]. Advanced understanding of the *S*-haplotype diversity, including dominance relationships between the haplotypes [10], is important for the efficient production of high-quality F$_1$-hybrid seed in *Brassica* crops.

In addition to SI, we reported an interesting incompatibility relationship between Turkish and Japanese populations of *B. rapa* [11,12]. Pollen of the Turkish line was rejected on the stigma of the Japanese line, although crossing in the reverse direction showed compatibility. This cross-incompatibility operating in one direction, unilateral incompatibility

(UI) occurred within species, in contrast to the UI that is known to occur in interspecies crossing [13,14]. Our molecular genetic studies of intraspecies UI in *B. rapa* revealed that it was controlled by the stigma-expressed gene, *stigmatic unilateral incompatibility 1, SUI1*, encoding an SRK-like receptor kinase and the pollen-expressed gene, *pollen unilateral incompatibility 1, PUI1*, encoding an SP11-like small cysteine-rich ligand. *SUI1* and *PUI1* are tightly linked and are considered to originate from a duplication event of the *SRK-SP11* region in *Brassica* [12]. The *S* locus is located on chromosome A07, while the *UI* locus (containing *SUI1* and *PUI1*) is on chromosome A04 of *B. rapa* [12]. From our further analysis of genetic diversity and distribution of the *PUI1* and *SUI1* genes in *B. rapa*, a functional *PUI1-1* allele was found only in the Turkish lines and not in the Japanese lines, while the three functional *SUI1* alleles (*SUI1-1, -2,* and *-3*) were found in Japanese wild populations and some cultivated varieties. Thus, loss of function of *SUI1* in Turkish lines and *PUI1* in Japanese lines might have resulted in the unidirectional pollen-stigma incompatibility in *B. rapa* [12].

The physiological pollen-rejection phenotype of the intraspecies UI is similar to that of SI and is consistent with the involvement of *M*-locus protein kinase (MLPK) in UI, which may function in SRK-mediated SI signal transduction [15,16]. It is noteworthy that the incompatibility response of UI is almost as strong as in the rigid SI phenotype in *B. rapa*. Thus, UI may have an effect on the SI-dependent breeding process in *B. rapa*. In this study, we extensively analyzed *SUI1* and *PUI1* alleles in Japanese cultivated lines of Chinese cabbage (*Brassica rapa* var. *pekinensis*). The results presented here give new insight into the historical relationship between UI and the breeding system of Chinese cabbage in Japan.

2. Results

2.1. Cultivars of Chinese Cabbage Produced by Japanese Seed Companies

The UI phenotype observed on the stigma (stigma-side UI phenotype) was originally identified in the Japanese commercial hybrid variety 'Osome' of Japanese mustard spinach, Komatsuna (*B. rapa* var. *perviridis*), from the Takii seed company [11]. To understand the role of *SUI1* in Japanese *B. rapa* cultivars, here we examined 52 commercial cultivars of Chinese cabbage (*B. rapa* var. *pekinensis*) from 16 Japanese seed companies (listed in Table 1) to determine their *SUI1* and *PUI1* alleles in addition to their stigma-side UI phenotype. All the cultivars used in this study, except 'Kashinhakusai' (#8), are F$_1$ hybrids. Because functional *SUI1* alleles behave as dominant over nonfunctional alleles [11], they can be analyzed to predict the UI phenotype on the stigma side of hybrid varieties.

Table 1. UI phenotype and genotype of Japanese cultivars of Chinese cabbage.

Sample Number	Seed Company	Cultivar	Stigma-Side UI Phenotype	Genotype	
				SUI1	PUI1
#1	Tokita Seed Co., Ltd.	Mainoumi	UI	SUI1-2/SUI1-11	pui1-3/pui1-4/pui1-6
#2	Takii & Co., Ltd.	Puchihiri	UI	SUI1-2	pui1-3/pui1-4
#3	Takii & Co., Ltd.	Kigokoro 75	UI	SUI1-2	pui1-3/pui1-4
#4	Sakata Seed Corp.	Kimikomachi	UI	SUI1-2	pui1-3/pui1-4
#5	Takii & Co., Ltd.	Chihiri 70	UC	sui1-t10	pui1-3/pui1-4
#6	Takii & Co., Ltd.	Banki	UI	SUI1-2	nd
#7	Watanabe Seed Co., Ltd.	Matsushima shin2gou	UI	nd	pui1-3/pui1-4
#8	Noguchi Seed Co.	Kashinhakusai	UI	SUI1-1	pui1-3/pui1-4
#9	Watanabe Seed Co., Ltd.	Menkoi	UI	SUI1-2/SUI1-11	pui1-3/pui1-4/pui1-6
#10	Ishii Seed Growers Co., Ltd.	Kinami 90	UI	nd	nd
#11	Kaneko Seed Co., Ltd.	Kougetsu 77	UI	SUI1-2	pui1-3/pui1-4
#14	Kaneko Seed Co., Ltd.	Eiki	UC	nd	pui1-3/pui1-4/pui1-6
#16	Kaneko Seed Co., Ltd.	Moeki	UI	nd	pui1-3/pui1-4
#17	Kaneko Seed Co., Ltd.	Kasumihakusai	UC	SUI1-2/SUI1-12	pui1-3/pui1-4/pui1-6
#18	Kaneko Seed Co., Ltd.	Shouki	UI	nd	pui1-3/pui1-4
#19	Watanabe Seed Co., Ltd.	Strong CR	UI	SUI1-2	pui1-3/pui1-4
#20	Watanabe Seed Co., Ltd.	Aiki	UI	nd	pui1-3/pui1-4/pui1-6

Table 1. Cont.

Sample Number	Seed Company	Cultivar	Stigma-Side UI Phenotype	Genotype SUI1	Genotype PUI1
#23	Nozaki Saishujo Ltd.	Maiko	nd	nd	pui1-3/pui1-4
#24	Nozaki Saishujo Ltd.	Chi China	UI	nd	nd
#25	Nozaki Saishujo Ltd.	Eisyun	nd	nd	pui1-3/pui1-4/pui1-6
#27	Nozaki Saishujo Ltd.	Retasai	UI	nd	pui1-3/pui1-4/pui1-6
#33	Marutane Seed Co., Ltd.	Chikara	UI	SUI1-2	pui1-3/pui1-4
#35	Yamato Noen Co., Ltd.	Kiyorokobi	UI	SUI1-2	pui1-3/pui1-4
#41	Watanabe Seed Co., Ltd.	Kiai 65	UI	nd	pui1-3/pui1-4
#45	Kaneko Seed Co., Ltd.	Gokui	UC	nd	pui1-3/pui1-4/pui1-6
#47	Kaneko Seed Co., Ltd.	Taibyou nozomi 60	UI	nd	pui1-3/pui1-4
#49	Mikado Kyowa Seed Co., Ltd.	CR Ouken	UI	SUI1-2	pui1-3/pui1-4
#50	Mikado Kyowa Co., Ltd.	Hakuei hakusai	UC	SUI1-2/SUI1-10	pui1-3/pui1-4
#51	Mikado Kyowa Co., Ltd.	Senki	nd	nd	pui1-3/pui1-4/pui1-6
#53	Sakata Seed Corp.	Saiki	nd	nd	pui1-3/pui1-4/pui1-6
#55	Sakata Seed Corp.	Yumebuki	UI	nd	pui1-3/pui1-4/pui1-6
#57	Takayama Seed Co., Ltd.	Gokigen	UI	nd	pui1-3/pui1-4
#58	Ishii Seed Growers Co., Ltd.	CR Seiga 65	UI	SUI1-2	pui1-3/pui1-4
#62	Takii & Co., Ltd.	Oushou	UI	nd	nd
#63	Takii & Co., Ltd.	Musou	UI	SUI1-2	pui1-3/pui1-4
#64	Takii & Co., Ltd.	Senshou	UI	nd	pui1-3/pui1-4
#65	Takii & Co., Ltd.	Kinshou	UI	nd	pui1-3/pui1-4
#74	Tohoku Seed Co., Ltd.	Daifuku	UI	SUI1-2	pui1-3/pui1-4
#75	Tohoku Seed Co., Ltd.	Daifuku75	UI	nd	pui1-3/pui1-4
#77	Tohoku Seed Co., Ltd.	Shinseiki	UI	nd	pui1-3/pui1-4
#80	Nanto Seed Co., Ltd.	CR Kinshachi 75	UI	SUI1-2	pui1-3/pui1-4
#83	Nanto Seed Co., Ltd.	Taibyou apolo 60	UC	SUI1-2/SUI1-10	pui1-3/pui1-4
#84	Nippon Norin Seed Co.	Kikumusume 65	UI	SUI1-2	pui1-3/pui1-4
#85	Nippon Norin Seed Co.	Kien75	UI	nd	pui1-3/pui1-4
#88	Nippon Norin Seed Co.	Super CR Shinrisou	UC	SUI1-2	pui1-3/pui1-4
#91	Takayama Seed Co., Ltd.	Kinkaku 65	UI	nd	pui1-3/pui1-4/pui1-6
#96	Tokita Seed Co., Ltd.	Haruhi	UI	nd	pui1-3/pui1-4/pui1-6
#97	Nanto Seed Co., Ltd.	Taiki 60	UI	nd	pui1-3/pui1-4
#101	Musashino Seed Co., Ltd.	Nanzan	UI	nd	pui1-3/pui1-4/pui1-6
#102	Watanabe Seed Co., Ltd.	Seitoku	nd	nd	pui1-3/pui1-4
#103	Watanabe Seed Co., Ltd.	Shunjuu	UI	nd	pui1-3/pui1-4
#104	Watanabe Seed Co., Ltd.	Kaname	UI	nd	pui1-3/pui1-4

nd, not determined.

2.2. Stigma-Side UI Phenotype Determined by Pollination Test

To verify the stigma-side UI phenotype of the Japanese cultivars of Chinese cabbage, stigmas of 47 cultivars were crossed with the pollen from the Turkish line ($S^{24}t$, $S^{40}t$, or $S^{21}t$) possessing *PUI1-1/PUI1-1* with crossing combinations of different *S*-haplotypes for discriminating the UI phenotype from the SI phenotype ($S^{21}t$ was produced for this study) [16]. Among the 47 cultivars, 85% (40 cultivars) had the incompatibility (UI) phenotype to the Turkish pollen (Table 1). Only seven cultivars, 'Chihiri 70' (#5), 'Eiki' (#14), 'Kasumihakusai' (#17), 'Gokui' (#45), 'Hakuei hakusai' (#50), 'Taibyou apolo 60' (#83), and 'Super CR Shinrisou' (#88), had the compatibility (UC) phenotype to the Turkish pollen (Table 1). Thus, the majority of the Chinese cabbage cultivars we tested have the ability to reject the *PUI1-1/PUI1-1* pollen, indicating that they possess functional *SUI1* allele(s).

2.3. The SUI1 Allele and Its Distribution

We isolated the full-length *SUI1* gene by polymerase chain reaction (PCR) amplification from the genomic DNA of each cultivar and determined its allele(s) by sequencing, as listed in Table 1. From 22 cultivars in which *SUI1* was sequenced, six alleles, including functional alleles (*SUI1-1* and *-2*), were identified. One cultivar, 'Kashinhakusai' (#8), with stigmatic UI phenotype, had the *SUI1-1* allele, which was originally isolated from Komatsuna variety 'Osome' [11,12]. This may be because, among the cultivars used in the present study, only 'Kashinhakusai' (#8) is not an F_1 hybrid, as described above. The 16 cultivars with stigmatic UI phenotype possessed the *SUI1-2* allele (Table 1), which has been found in wild *B. rapa* populations [11,12]. Three alleles encoding putative intact SUI1 proteins, *SUI1-10* (accession, LC641787), *SUI1-11* (accession, LC641786), and *SUI1-12* (accession, LC641785), and one allele encoding truncated protein, *sui1-t10* (accession, LC641784) were newly identified alleles in this study (Figure 1). Phylogenetic analysis with amino acid sequences revealed that *SUI1-11* and *SUI1-12* belonged to the same clade, and this was different from the clade with the functional SUI1s (SUI1-1, -2, and -3) and SUI1-10 (Figure 2), suggesting that *SUI1-11* and *SUI1-12* are nonfunctional alleles. Four out of seven stigmatic UC cultivars possessed *SUI1-10*, *-12*, or *sui1-t10*.

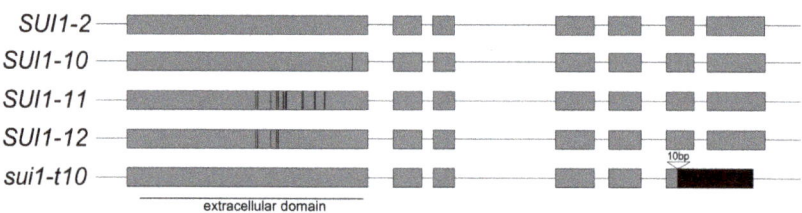

SUI1-10; C413Y

SUI1-11; A238E, H240Q, H265Y, R276Y, V279E, Q280R, W288S, P289E, L292R, L294E, E295D, R322H, I326L, L344I, R346N, R363H, V364D

SUI1-12; A238E, H240Q, H265Y, R276Y, V279E, Q280R

Figure 1. Schematic representation of the *SUI1* genomic sequences in this study. The shaded boxes represent the protein coding regions. Positions of amino acid substitutions compared to *SUI1-2* are shown by bars and listed below. The extracellular domain (consisting of most of the 1st exon) is indicated. The position of the 10-bp deletion of *sui1-t10* is shown in the sixth exon.

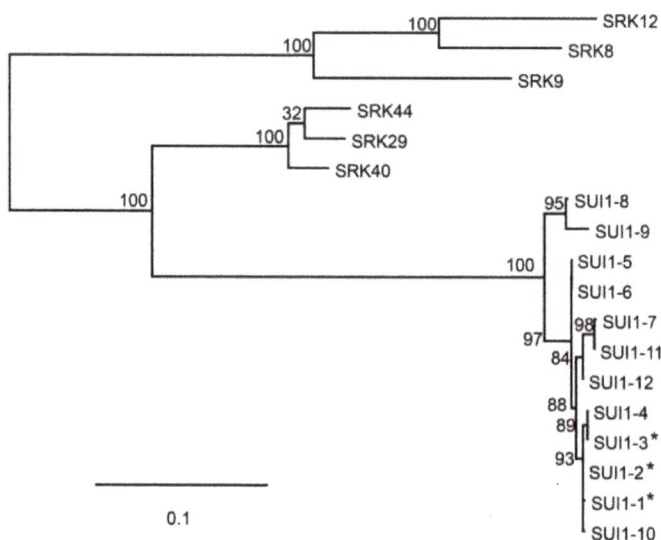

Figure 2. A maximum likelihood phylogenetic tree of SUI1s and SRKs in *B. rapa*. Branch support values from 100 bootstraps are indicated. Functional SUI1s that genetically interact with PUI1-1 are indicated by asterisks (*).

In the case of the *SUI1-10* allele, found in cultivars 'Hakuei hakusai' (#50) and 'Taibyou apolo 60′ (#83), a single base substitution at codon 413 (changing the residue from cysteine to tyrosine) was present at the C-terminus of the extracellular domain (Figure 1). Both cultivars possessing the *SUI1-10* allele showed stigmatic UC phenotype, despite being heterozygous for the functional *SUI1-2* allele (Table 1), indicating that there was a dominant negative effect of *SUI1-10* toward *SUI1-2* (as described below in detail).

On the other hand, the *SUI1-11* allele had 17 amino acid changes in the extracellular domain, and cultivars 'Mainoumi' (#1) and 'Menkoi' (#9) with *SUI1-2/SUI1-11* heterozygote showed the stigmatic UI phenotype (Figure 1, Table 1). Even if *SUI1-11* is nonfunctional, as expected, the stigmatic UI phenotype is consistent with the dominance of *SUI1-2* over *SUI1-11*.

The *SUI1-12* allele had six amino acid changes in the extracellular domain. The cultivar 'Kasumihakusai' (#17) had *SUI1-2* and *SUI1-12* alleles as a heterozygote, and it showed the stigmatic UC phenotype (Figure 1, Table 1). It is also possible that *SUI1-12* might show a dominant negative effect to *SUI1-2* in 'Kasumihakusai' (#17), as in the case of *SUI1-10* in 'Hakuei hakusai' (#50) and 'Taibyou apolo 60′ (#83).

'Chihiri 70′ (#5) possessed the truncated *sui1-t10* allele (Figure 1). All the 15 *SUI1* clones of 'Chihiri 70′ (#5) isolated from two independent PCR amplifications were *sui1-t10*, suggesting that 'Chihiri 70′ (#5) is homozygous for *sui1-t10*, which is consistent with its stigmatic UC phenotype. The sequence of the extracellular domain of *sui1-t10* was perfectly matched with *SUI1-1* and *SUI1-2* functional alleles, but there was a 10-bp deletion in the sixth exon, as in *sui1-t4, sui1-t5,* and *sui1-t6*, which results in a frameshift and creates a premature termination codon [12].

2.4. The PUI1 Allele and Its Distribution

To examine the *PUI1* alleles of 48 cultivars of Chinese cabbage, we cloned the PCR fragments of the full-length *PUI1* and determined their sequences (Table 1, see Materials and Methods section). In these Japanese cultivars, we found three nonfunctional alleles (*pui1-3, -4,* and *-6*), which have been reported previously [12]. Out of the 48 cultivars, 34 possessed both *pui1-3* and *pui1-4*, and 14 possessed all three alleles (Table 1). The existence of three alleles in an individual plant indicates the possibility of duplication of *PUI1*. To verify this duplication, we first propagated the self-pollinated progeny of 'Super CR Shinrisou' (#88; *pui1-3/pui1-4*) and determined the *PUI1* genotype of the 22 segregants using a direct sequencing method. It was found that all the segregants exhibited both *pui1-3* and *pui1-4*, suggesting that the *pui1-3* and *pui1-4* genes were linked and homozygous in this progeny (Table 2). Next, we propagated the self-pollinated progeny of 'Gokui' (#45; *pui1-3/pui1-4/pui1-6*) and determined the *PUI1* genotype of the 32 segregants using a PCR-restriction fragment length polymorphism (RFLP) method. It was found that all individuals possessed *pui1-3, pui1-4,* and *pui1-6*, suggesting that the three *PUI1* genes (*pui1-3/pui1-4/pui1-6*) were linked and homozygous in this progeny (Table 2). Furthermore, a similar PCR-RFLP experiment was performed using 'Nanzan' (#101; *pui1-3/pui1-4/pui1-6*) selfed progeny (Table 2, Table S1). Interestingly, the self-pollinated population (78 plants) of 'Nanzan' segregated to *pui1-3/pui1-4* and *pui1-3/pui1-4/pui1-6* plants. Their segregation ratio was 17:61 (1:3; chi-square test, $\chi^2 = 0.43, p > 0.05$) fit for a simple Mendelian inheritance. The result indicates that 'Nanzan' (#101) is a heterozygote of duplicated (*pui1-3/pui1-4*) and triplicated (*pui1-3/pui1-4/pui1-6*) *PUI1* genes in the *B. rapa* genome. Thus, duplication and/or triplication of nonfunctional *PUI1* genes had occurred at the *UI* locus region in Japanese *B. rapa* cultivars.

Table 2. Segregation analysis of *PUI1* allele in the selfed progeny of #45, #88, and #101.

Sample Number	Cultivar	n	PUI1 Genotype Detected	
			pui1-3/pui1-4	pui1-3/pui1-4/pui1-6
#45	Gokui	32	0	32
#88	Super CR Shinrisou	22	22	-*
#101	Nanzan	78	17	61

*, 'Super CR Shinrisou' (#88) does not have *pui1-6* allele.

2.5. Genetic Segregation Analysis of the Dominant Negative Effect of SUI1-10

As described above, 'Hakuei hakusai' (#50) and 'Taibyou apolo 60' (#83), possessing the *SUI1-2/SUI1-10* genotype, exhibited stigma-side UC phenotype (i.e., accepting the Turkish *PUI1-1/PUI1-1* pollen), even though they have a functional *SUI1-2* allele. To confirm this dominant negative effect of *SUI1-10*, we performed a genetic analysis of 'Taibyou apolo 60' (#83).

We produced self-pollinated progeny of 'Taibyou apolo 60' (#83-S_1 progeny) and determined their stigma-side UI phenotype and *SUI1* genotype (Table 3). Stigma-side UI phenotypes of this progeny were determined by test cross-pollination using homozygous plants ($S^{24}t$) as the pollen donor. The *SUI1-2* and *SUI1-10* alleles were discriminated by direct-sequencing detection of a single nucleotide polymorphism at codon 413 and were segregated in the #83-S_1 progeny; three of eleven plants showed stigma-side UI, and the others were stigma-side UC. Stigma of thee *SUI1-2/SUI1-2* homozygous plants were incompatible to the $S^{24}t$ pollen (UI), and five *SUI1-2/SUI1-10* heterozygous, and three *SUI1-10/SUI1-10* homozygous individuals showed compatible pollen tube penetration with the $S^{24}t$ pollen (UC), indicating that *SUI1-10* is nonfunctional and has a dominant negative effect to the functional *SUI1-2*.

Table 3. Segregation analysis of *SUI1* allele in the selfed progeny of #83.

Population	SUI1 Genotype	n	Stigma-Side UI Phenotype	
			UI	UC
#83-S_1	SUI1-2/SUI1-2	3	3	0
	SUI1-2/SUI1-10	5	0	5
	SUI1-10/SUI1-10	3	0	3
#83-S_2	SUI1-2/SUI1-2	23	23	0
	SUI1-2/SUI1-10	41	0	41
	SUI1-10/SUI1-10	16	0	16

For further confirmation of this effect, the #83-S_2 population with a higher number of plants was produced by self-bud pollination of the #83-S_1 *SUI1-2/SUI1-10* heterozygous plants. In the #83-S_2 population, *SUI1* genotypes segregated as expected; for genotypes *SUI1-2/SUI1-2*: *SUI1-2/SUI1-10*: *SUI1-10/SUI1-10* the observed ratio was 23:41:16 (1:2:1; chi-square test, $\chi^2 = 1.27$, $p > 0.05$, $df = 2$, Table 3, Table S2). The stigma-side UC phenotype and *SUI1-10* genotype of the #83-S_2 population showed perfect linkage in the 80 plants (Table 3). Thus, it was concluded that the nonfunctional *SUI1-10* does show a dominant negative effect on the functional *SUI1-2*.

To verify if this effect is observed with the other functional allele, we produced *SUI1-3/SUI1-10* heterozygous plants by a cross between *SUI1-3/SUI1-3* [11,12] and *SUI1-10/SUI1-10* plants selected from the #83-S_2 population. Stigmas of *SUI1-3/SUI1-10* heterozygous plants were compatible (UC) with *PUI1-1/PUI1-1* pollen from the $S^{24}t$ and also $S^{40}t$ lines, indicating that *SUI1-10* also shows a dominant negative effect on the functional *SUI1-3*.

3. Discussion

Highly controlled pollen-stigma incompatibility is important for F_1 hybrid production of *Brassica* cultivars. The molecular mechanism of SI in *Brassica* has been studied for many years and is used in F_1 breeding. The recently discovered UI system, regulated by *SUI1* and *PUI1*, can potentially provide another mechanism to control pollination in *B. rapa*. Therefore, determination of the *UI* genotype is considered as important as the SI genotype in the breeding of this major Japanese vegetable, Chinese cabbage. In this study, we determined the *SUI1* and *PUI1* allelic diversity of 22 and 48 cultivars, respectively, of Chinese cabbage in Japan. In addition, we confirmed the stigma-side UI phenotype of 47 cultivars. This revealed that most of the cultivars showed a stigma-side UI phenotype with a functional *SUI1* allele (*SUI1-2*), whereas no functional *PUI1* allele (*PUI1-1*) was found. We also searched the re-sequence data of *B. rapa* lines that are stocked at Chungnam National University and found a functional *SUI1-2* allele in a South Korean population (data not shown). The fact that functional *SUI1* alleles are present in Japanese and South Korean cultivars should be taken into consideration in breeding programs for *B. rapa*. UI may be beneficial as the additional incompatibility, which could be used in breeding programs by the introduction of *PUI1-1* to the pollen donor.

To the best of our knowledge, there is no report that traits important for Chinese cabbage are mapped to flanking regions of the *UI* locus in chromosome A04. Thus, for an unknown reason, the functional *SUI1-2* has been selected, and its sequence has been conserved during the breeding of Chinese cabbage cultivars in Japan. It would be interesting to investigate whether *SUI1* itself strengthens SI and thus increases the efficiency of F_1 seed production.

In our previous study, we isolated nine intact alleles of *SUI1* and showed that *SUI1-1*, *SUI1-2*, and *SUI1-3* are incompatible with *PUI1-1*/*PUI1-1* pollen [12]. *SUI1-1* was originally isolated from a Japanese commercial hybrid variety of Komatsuna (*B. rapa* var. *perviridis*), and *SUI1-2* and *SUI1-3* were found in Japanese wild populations of *B. rapa* [12]. In the current study, we isolated three novel intact *SUI1* alleles; one (*SUI1-10*) belongs to the functional clade (with *SUI1-1*, *SUI1-2*, and *SUI1-3*) and the other two alleles (*SUI1-11* and *SUI1-12*) belong to the nonfunctional clade (Figure 2). The fact that *SUI1-10*/*SUI1-10* homozygote is stigmatic UC indicates that *SUI1-10* is a nonfunctional allele (Table 3). The Cys-413 residue of *SUI1-2* is the last of the 12 highly conserved cysteine residues in the *SUI1* extracellular domain and is located within the PAN_APPLE domain, which is the C terminal region of the extracellular receptor region. It has been clarified that homodimerization of SRK in Brassicaceae is essential for ligand interaction [17]. The PAN_APPLE domain of SRK has been shown to be important for ligand-independent dimer formation of SRKs and is responsible for correct intracellular trafficking [18–21]. It has been reported that the last Cys residue of SRK is predicted to form an intramolecular disulfide bond [20,21]. Thus, although the *SUI1-10* sequence is similar to the functional *SUI1-2*, the C413Y mutation of *SUI1-10* might cause structural disruption of *SUI1* and breakdown of incompatibility through unusual dimer formation.

A feature of the sporophytic regulation of SI is the dominance relationship between *S*-haplotypes [10,22,23]. The molecular mechanism of the pollen-side dominance relationship has been well studied and revealed that mono-allelic gene expression of the dominant *SP11* haplotype is controlled by small RNA-based epigenetic regulation [24–26]. On the stigma side, there is a complex allelic interaction that is as yet unexplained [10]. It was presumed that the SRK protein itself determines the dominance relationship rather than differences in *SRK* gene expression [23], and Naithani et al. [18] noted that the stigma-side dominance relationship may result from an increased tendency for heterodimer formation in some SRK pairs [18]. On the other hand, the existence of dominant negative alleles of receptor kinases that function as receptor complexes in many situations during plant development is widely known [27–29]. In most of these, the formation of a receptor complex with abnormal receptor proteins or receptor-related proteins encoded by dominant negative alleles causes disruption of signaling pathways. Thus, one possible explanation for the dominant negative effect of *SUI1-10* may be an increase of *SUI1-2/SUI1-10* heterodimer on the stigma surface and competitive inhibition of the interaction with the *PUI1* ligand. We also found a dominant negative effect of *SUI1-10* to *SUI1-3*, which has four aa substitutions (R322H, I326L, R363H, and V364D) compared to the extracellular domain of *SUI1-2*, suggesting that these four residues are not important for the effect.

In this study, it was found that the *PUI1* gene of Japanese cultivars of Chinese cabbage showed very low diversity. Among six *PUI1* alleles, of which only *PUI1-1* from a Turkish strain can induce UI [12], only two patterns of genotype (*pui1-3/pui1-4* or *pui1-3/pui1-4/pui1-6*) were observed, and no cultivars with a functional *PUI1-1* allele could be found. Interestingly, the *pui1-3/pui1-4* genotype might consist of two linked *pui1-3* and *pui1-4* genes (Figure S1). Similarly, the *pui1-3/pui1-4/pui1-6* genotype might consist of three linked *pui1-3*, *pui1-4*, and *pui1-6* genes (Figure S1). Such duplication and triplication of nonfunctional *PUI1* have complicated the *UI* locus region. Although such *PUI1* duplication or triplication cannot be found in the reference genome information of *B. rapa* inbred line Chiifu (*B. rapa* reference genome version 3.0, https://brassicadb.cn, accessed on 1 April 2021), de novo genomic sequence assembly of these Chinese cabbage cultivars using next-generation sequencing technology, including long-read sequencing, would provide new insights into the genomic structure of the *UI* locus [30]. In fact, we can find the two duplicated *PUI1* genes on the *UI* locus of the genome sequence of *B. rapa* Z1(version 1.0, https://brassicadb.cn, accessed on 19 October 2021, Figure S2) [31].

Further analysis of the genetic diversity of the *UI* locus in *B. rapa* other than Chinese cabbage (subsp. *pekinensis*), such as turnips (subsp. *rapa*), leafy *Brassica* crops (subsp. *chinensis*, *periridis*), and field mustard (subsp. *oleifera*) will not only contribute to the discovery of novel alleles but also provide new insights into the genomic structure of the pollen-side factor and the dominant recessive interaction of the stigma-side factor. It will also be interesting to determine whether the *UI* locus has a multi-allelic structure like the *S* locus.

4. Materials and Methods

4.1. Plant Material

The plant material consisted of 52 commercial cultivars of Chinese cabbage, *B. rapa* ssp. *pekinensis* (Table 1). All except one, 'Kashinhakusai,' were F_1 hybrid cultivars. To produce self-pollinated progeny, bud pollination was performed. Petals and stamens were removed from a young flower bud (2–4 d before flowering), and the immature pistil was pollinated. The pollinated pistil was then covered with a paper bag until the seed was harvested. Plant materials were vernalized at 4 °C for 4 weeks in a refrigerator and then grown in a greenhouse.

4.2. Test Pollination

Flower buds were cut at the peduncle and pollinated. After pollination, they were stood on 1% solid agar for about 24 h under room conditions. Then, pistils of the pollinated flowers were softened in 1N NaOH for 1 h at 60 °C and stained with basic aniline blue (0.1 M K_3PO_4, 0.1% aniline blue). Samples were mounted in 50% glycerol on slides and observed by UV fluorescence microscopy (Figure S3) [32]. At least three flowers were used from each cross combination, and observations were generally replicated at least three times on different dates for each cross combination. For the determination of the stigma-side UI phenotype, *PUI1-1/PUI1-1* homozygous plants ($S^{24}t$, $S^{40}t$, and $S^{21}t$) were used as the pollen donor in test pollinations ($S^{21}t$ was produced for this study) [16].

4.3. Cloning, Sequencing, and Genotyping of SUI1 and PUI1 Alleles

Total DNA was extracted from young leaf tissue of *B. rapa* by the procedure of Murray and Thompson (1980) or using a DNeasy plant mini kit (Qiagen) [33]. For molecular cloning of full-length *SUI1* and *PUI1* genes, genomic PCR was performed using KOD-Plus-Neo DNA polymerase (TOYOBO) according to the manufacturer's instructions. PCR primers

SUI1cDNA_F3 and SUI1_gR2 for *SUI1* and PCP-like1-F1 and PCP-like1-R1 for *PUI1* were used (Table S3). All amplified fragments were detected as a single band in the gel electrophoresis. PCR products were modified by adding 3'-A overhangs using A-attachment mix (TOYOBO) and cloned into a vector, pTAC-2, using DynaExpress TA PCR Cloning kit (Biodynamics). The nucleotide sequence was determined with a 3500 or 310 Genetic Analyzer using Big Dye Terminator version 3.1 or 1.1 Cycle Sequencing Kit (Applied Biosystems); in the case of *SUI1*, the *SUI1*-specific sequencing primers, SUIcDNA_F3, SUI_gR2, SUIinter_cF1, SUIinter_cF2, SUIinter_cF3, SUIinter_GF1, SUI1inter_cF4, and SUIinter_cF5 (Figure 1 and Table S3), were used. GENETYX version 13 software package (GENETYX Corp.) was used for the sequence comparison and alignment. For the segregation analysis, we determined the genotype of *SUI1* and *PUI1* alleles by direct sequencing of PCR products. *SUI1-1* and *SUI1-10* alleles were amplified using primers SUI1_2-10typeSDF and SUI1_2-10typeSDR (Table S3). Each *PUI1* allele was amplified using the primer pair for the *PUI1* second exon region, PUI1-3.4.6-F, and PUI1-3.4.6-R (Table S3, Figure S4). For discrimination of *PUI1* alleles by PCR-RFLP, amplified DNA fragments were cut by restriction enzyme (BamHI, SalI, or BsrI), followed by checking on an electrophoresed agarose gel (Figure S4). For the direct sequencing marker, amplified fragments were purified from the electrophoresed agarose gel and sequenced as described above.

4.4. Phylogenetic Analysis

Phylogenetic analysis was performed on the Phylogeny.fr platform (http://www.phylogeny.fr/, accessed on 21 October 2021) [34]. Full-length amino acid sequences were aligned with MUSCLE (version 3.7) configured for the highest accuracy. Accession numbers of SRKs and SUI1s are listed in Table S4. After alignment, ambiguous regions were removed with Gblocks (version 0.91b). The phylogenetic tree was reconstructed using the PhyML program (version 3.0 aLRT). The default substitution model was selected assuming an estimated proportion of invariant sites and 4 gamma-distributed rate categories to account for rate heterogeneity across sites. The reliability of internal branches was assessed using the bootstrapping method (100 bootstrap replicates). The tree was represented with TreeDyn (version 198.3).

Supplementary Materials: The following are available online at https://www.mdpi.com/article/10.3390/plants10112467/s1, Figure S1: Schematic model of duplicated and triplicated *PUI1* allele; Figure S2: Genomic organization of the *SUI1* and *PUI1* region of *B. rapa* Z1 (*yellow sarson*) identified from published genome sequence available at https://brassicadb.cn accessed on 29 September 2021 [31]; Figure S3: Representative results of test-pollination under UV fluorescence microscopy; Figure S4: Nucleotide sequence alignment of *PUI1* alleles. Table S1: *PUI1* genotype in the selfed progeny of #101, 'Nanzan'; Table S2: *SUI1* genotype and stigma-side UI phenotype of selfed progeny of #83; Table S3: Primers used in this study; Table S4: Accession number of *SUI1* and SRK sequences used in phylogenetic analysis.

Author Contributions: Y.T., Y.-P.L., S.T., G.S., and M.W. conceived and designed the experiments. Y.T., A.M., Y.H., H.X., Y.O., H.N., and S.H. performed the research and analyzed the data. Y.T., S.T., G.S., and M.W. wrote the paper, which was edited by all other authors. All authors have read and agreed to the published version of the manuscript.

Funding: This work was supported in part by MEXT KAKENHI (Grant Numbers 16H06470, 16H06464, and 16K21727 to MW), JSPS KAKENHI (Grant Numbers 19K05963 to YT; 16H06380, 21H05030 to ST; 20K05982 to GS; 16H04854, 16K15085, 17H00821, 18KT0048, 21H02162 to MW). Korea Institute of Planning and Evaluation for Technology in Food, Agriculture, and Forestry (IPET) through Golden Seed Project (project number: 213006-05-5-SB110 to YPL and SH), Ministry of Agriculture, Food and Rural Affairs (MAFRA), Ministry of Oceans and Fisheries (MOF), Rural Development Administration (RDA) and Korea Forest Services (KFS) to YPL and SH. This research was also supported by the Japan Advanced Plant Science Network to MW.

Data Availability Statement: The available data are presented in the manuscript.

Acknowledgments: The authors thank Masuko-Suzuki Hiromi, Kana Ito, Nono Sugawara, Megumi Ito, Keita Yamaki, Tai Takemoto and Misono Sasaki (Tohoku University) for technical assistance.

Conflicts of Interest: The authors declare no conflict of interest.

References

1. Bateman, A.J. Self-incompatibility systems in angiosperms. III. Cruciferae. *Heredity* **1955**, *9*, 53–68. [CrossRef]
2. Suzuki, G.; Kai, N.; Hirose, T.; Fukui, K.; Nishio, T.; Takayama, S.; Isogai, A.; Watanabe, M.; Hinata, K. Genomic organization of the *S* locus: Identification and characterization of genes in *SLG/SRK* region of S^9 haplotype of *Brassica campestris* (syn. *rapa*). *Genetics* **1999**, *153*, 391–400. [CrossRef] [PubMed]
3. Takasaki, T.; Hatakeyama, K.; Suzuki, G.; Watanabe, M.; Isogai, A.; Hinata, K. The *S* receptor kinase determines self-incompatibility in *Brassica* stigma. *Nature* **2000**, *403*, 913–916. [CrossRef] [PubMed]
4. Schopfer, C.R.; Nasrallah, M.E.; Nasrallah, J.B. The male determinant of self-incompatibility in *Brassica*. *Science* **1999**, *286*, 1697–1700. [CrossRef]
5. Takayama, S.; Shiba, H.; Iwano, M.; Shimosato, H.; Che, F.-S.; Kai, N.; Watanabe, M.; Suzuki, G.; Hinata, K.; Isogai, A. The pollen determinant of self-incompatibility in *Brassica campestris*. *Proc. Natl. Acad. Sci. USA* **2000**, *97*, 1920–1925. [CrossRef]
6. Takayama, S.; Shimosato, H.; Shiba, H.; Funato, M.; Che, F.-S.; Watanabe, M.; Iwano, M.; Isogai, A. Direct ligand-receptor complex interaction controls *Brassica* self-incompatibility. *Nature* **2001**, *413*, 534–538. [CrossRef]
7. Nou, I.S.; Watanabe, M.; Isogai, A.; Shiozawa, H.; Suzuki, A.; Hinata, K. Variation of *S*-alleles and *S*-glycoproteins in a naturalized population of self-incompatible *Brassica campestris* L. *Jpn. J. Genet.* **1991**, *66*, 227–239. [CrossRef]
8. Nou, I.S.; Watanabe, M.; Isuzugawa, K.; Isogai, A.; Hinata, K. Isolation of *S*-alleles from a wild population of *Brassica campestris* L. at Balcesme, Turkey and their characterization by *S*-glycoproteins. *Sex. Plant Reprod.* **1993**, *6*, 71–78. [CrossRef]
9. Nou, I.S.; Watanabe, M.; Isogai, A.; Hinata, K. Comparison of *S*-alleles and *S*-glycoproteins between two wild populations of *Brassica campestris* in Turkey and Japan. *Sex. Plant Reprod.* **1993**, *6*, 79–86. [CrossRef]
10. Hatakeyama, K.; Watanabe, M.; Takasaki, T.; Ojima, K.; Hinata, K. Dominance relationships between *S*-alleles in self-incompatible *Brassica campestris* L. *Heredity* **1998**, *80*, 241–247. [CrossRef]
11. Takada, Y.; Nakanowatari, T.; Sato, J.; Hatakeyama, K.; Kakizaki, T.; Ito, A.; Suzuki, G.; Shiba, H.; Takayama, S.; Isogai, A.; et al. Genetic analysis of novel intra-species unilateral incompatibility in *Brassica rapa* (syn. *campestris*) L. *Sex. Plant Reprod.* **2005**, *17*, 211–217. [CrossRef]
12. Takada, Y.; Murase, K.; Shimosato-Asano, H.; Sato, T.; Nakanishi, H.; Suwabe, K.; Shimizu, K.K.; Lim, Y.P.; Takayama, S.; Suzuki, G.; et al. Duplicated pollen-pistil recognition loci control intraspecific unilateral incompatibility in *Brassica rapa*. *Nat. Plants* **2017**, *3*, 17096. [CrossRef]
13. Lewis, D.; Crowe, L.K. Unilateral incompatibility in flowering plants. *Heredity* **1958**, *12*, 233–256. [CrossRef]
14. Hiscock, S.J.; Dickinson, H.G. Unilateral incompatibility within the Brassicaceae: Further evidence for the involvement of the self-incompatibility (S)-locus. *Theor. Appl. Genet.* **1993**, *86*, 744–753. [CrossRef]
15. Murase, K.; Shiba, H.; Iwano, M.; Che, F.S.; Watanabe, M.; Isogai, A.; Takayama, S. A membrane-anchored protein kinase involved in *Brassica* self-incompatibility signaling. *Science* **2004**, *303*, 1516–1519. [CrossRef]
16. Takada, Y.; Sato, T.; Suzuki, G.; Shiba, H.; Takayama, S.; Watanabe, M. Involvement of MLPK pathway in intraspecies unilateral incompatibility regulated by a single locus with stigma and pollen factors. *G3 Genes Genomes Genet.* **2013**, *3*, 719–726. [CrossRef]
17. Shimosato, H.; Yokota, N.; Shiba, H.; Iwano, M.; Entani, T.; Che, F.S.; Watanabe, M.; Isogai, A.; Takayama, S. Characterization of the SP11/SCR high-affinity binding site involved in self/nonself recognition in *Brassica* self-incompatibility. *Plant Cell* **2007**, *19*, 107–117. [CrossRef]
18. Naithani, S.; Chookajorn, T.; Ripoll, D.R.; Nasrallah, J.B. Structural modules for receptor dimerization in the S-locus receptor kinase extracellular domain. *Proc. Natl. Acad. Sci. USA* **2007**, *104*, 12211–12216. [CrossRef]
19. Yamamoto, M.; Tantikanjana, T.; Nishio, T.; Nasrallah, M.E.; Nasrallah, J.B. Site-specific N-glycosylation of the S-locus receptor kinase and its role in the self-incompatibility response of the Brassicaceae. *Plant Cell* **2014**, *26*, 4749–4762. [CrossRef]
20. Ma, R.; Han, Z.; Hu, Z.; Lin, G.; Gong, X.; Zhang, H.; Nasrallah, J.B.; Chai, J. Structural basis for specific self-incompatibility response in *Brassica*. *Cell Res.* **2016**, *26*, 1320–1329. [CrossRef]
21. Murase, K.; Moriwaki, Y.; Mori, T.; Liu, X.; Masaka, C.; Takada, Y.; Maesaki, R.; Mishima, M.; Fujii, S.; Hirano, Y.; et al. Mechanism of self/nonself-discrimination in *Brassica* self-incompatibility. *Nat. Commun.* **2020**, *11*, 4916. [CrossRef]
22. Fujii, S.; Takayama, S. Multilayered dominance hierarchy in plant self-incompatibility. *Plant Reprod.* **2018**, *31*, 15–19. [CrossRef]
23. Hatakeyama, K.; Takasaki, T.; Suzuki, G.; Nishio, T.; Watanabe, M.; Isogai, A.; Hinata, K. The *S* receptor kinase gene determines dominance relationships in stigma expression of self-incompatibility in *Brassica*. *Plant J.* **2001**, *26*, 69–76. [CrossRef]
24. Shiba, H.; Iwano, M.; Entani, T.; Ishimoto, K.; Shimosato, H.; Che, F.S.; Satta, Y.; Ito, A.; Takada, Y.; Watanabe, M.; et al. The dominance of alleles controlling self-incompatibility in *Brassica* pollen is regulated at the RNA level. *Plant Cell* **2002**, *14*, 491–504. [CrossRef] [PubMed]
25. Tarutani, Y.; Shiba, H.; Iwano, M.; Kakizaki, T.; Suzuki, G.; Watanabe, M.; Isogai, A.; Takayama, S. *Trans*-acting small RNA determines dominance relationships in *Brassica* self-incompatibility. *Nature* **2010**, *466*, 983–986. [CrossRef]

26. Yasuda, S.; Wada, Y.; Kakizaki, T.; Tarutani, Y.; Miura-Uno, E.; Murase, K.; Fujii, S.; Hioki, T.; Shimoda, T.; Takada, Y.; et al. A complex dominance hierarchy is controlled by polymorphism of small RNAs and their targets. *Nat. Plants* **2016**, *3*, 16206. [CrossRef] [PubMed]
27. Shpak, E.D.; Lakeman, M.B.; Torii, K.U. Dominant-negative receptor uncovers redundancy in the *Arabidopsis* ERECTA leucine-rich repeat receptor-like kinase signaling pathway that regulates organ shape. *Plant Cell* **2003**, *15*, 1095–1110. [CrossRef] [PubMed]
28. Diévart, A.; Dalal, M.; Tax, F.E.; Lacey, A.D.; Huttly, A.; Li, J.; Clark, S.E. CLAVATA1 dominant-negative alleles reveal functional overlap between multiple receptor kinases that regulate meristem and organ development. *Plant Cell* **2003**, *15*, 1198–1211. [CrossRef]
29. Wang, X.; Li, X.; Meisenhelder, J.; Hunter, T.; Yoshida, S.; Asami, T.; Chory, J. Autoregulation and homodimerization are involved in the activation of the plant steroid receptor BRI1. *Dev. Cell* **2005**, *8*, 855–865. [CrossRef]
30. Zhang, L.; Cai, X.; Wu, J.; Liu, M.; Grob, S.; Cheng, F.; Liang, J.; Cai, C.; Liu, Z.; Liu, B.; et al. Improved *Brassica rapa* reference genome by single-molecule sequencing and chromosome conformation capture technologies. *Hortic. Res.* **2018**, *5*, 50. [CrossRef]
31. Belser, C.; Istace, B.; Denis, E.; Dubarry, M.; Baurens, F.C.; Falentin, C.; Genete, M.; Berrabah, W.; Chèvre, A.M.; Deloume, R.; et al. Chromosome-scale assemblies of plant genomes using nanopore long reads and optical maps. *Nat. Plants* **2018**, *4*, 879–887. [CrossRef]
32. Kho, Y.O.; Bear, J. Observing pollen tubes by means of fluorescence. *Euphytica* **1968**, *17*, 298–302. [CrossRef]
33. Murray, M.G.; Thompson, W.F. Rapid isolation of high molecular weight plant DNA. *Nucl. Acids Res.* **1980**, *8*, 4321–4325. [CrossRef]
34. Dereeper, A.; Guignon, V.; Blanc, G.; Audic, S.; Buffet, S.; Chevenet, F.; Dufayard, J.F.; Guindon, S.; Lefort, V.; Lescot, M.; et al. Phylogeny.fr: Robust phylogenetic analysis for the non-specialist. *Nucl. Acids Res.* **2008**, *36*, W465–W469. [CrossRef]

MDPI
St. Alban-Anlage 66
4052 Basel
Switzerland
Tel. +41 61 683 77 34
Fax +41 61 302 89 18
www.mdpi.com

Plants Editorial Office
E-mail: plants@mdpi.com
www.mdpi.com/journal/plants

www.ingramcontent.com/pod-product-compliance
Lightning Source LLC
LaVergne TN
LVHW070043120526
838202LV00101B/422